PREFACE

1. Scope

This publication establishes joint doctrine for the Armed Forces of the United States involved in or supporting foreign internal defense (FID). It discusses how joint operations, involving the application of all instruments of national power, support host nation efforts to build capability and capacity to free and protect its society from subversion, lawlessness, and insurgency.

2. Purpose

This publication has been prepared under the direction of the Chairman of the Joint Chiefs of Staff. It sets forth joint doctrine to govern the joint activities and performance of the Armed Forces of the United States in operations and provides the doctrinal basis for interagency coordination and for US military involvement in multinational operations while conducting or supporting FID. It provides military guidance for the exercise of authority by combatant commanders and other joint force commanders (JFCs) and prescribes joint doctrine for operations, education, and training. It provides military guidance for use by the Armed Forces in preparing their appropriate plans. It is not the intent of this publication to restrict the authority of the JFC from organizing the force and executing the mission in a manner the JFC deems most appropriate to ensure unity of effort in the accomplishment of the overall objectives.

3. Application

a. Joint doctrine established in this publication applies to the joint staff, commanders of combatant commands, subunified commands, joint task forces, subordinate components of these commands, the Services, and defense agencies in support of joint operations.

b. The guidance in this publication is authoritative; as such, this doctrine will be followed except when, in the judgment of the commander, exceptional circumstances dictate otherwise. If conflicts arise between the contents of this publication and the contents of Service publications, this publication will take precedence for the activities of joint forces unless the Chairman of the Joint Chiefs of Staff, normally in coordination with the other members of the Joint Chiefs of Staff, has provided more current and specific guidance. Commanders of forces operating as part of a multinational (alliance or coalition) military command should follow multinational doctrine and procedures ratified by the United States. For doctrine and procedures not ratified by the United States, commanders should evaluate and follow the multinational command's doctrine and procedures, where applicable and consistent with US law, regulations, and doctrine.

For the Chairman of the Joint Chiefs of Staff:

LLOYD J. AUSTIN III
Lieutenant General, USA
Director, Joint Staff

SUMMARY OF CHANGES
REVISION OF JOINT PUBLICATION 3-07.1, DATED 30 APRIL 2004 - RENUMBERED AS JOINT PUBLICATION 3-22

• **Changes title from** *Joint Tactics, Techniques, and Procedures for Foreign Internal Defense* **to** *Foreign Internal Defense* **and the publication number from Joint Publication (JP) 3-07.1 to JP 3-22**

• **Adds discussion of foreign internal defense (FID) within the range of military operations**

• **Elevates the internal defense and development appendix to a chapter, explaining FID construct, functions, principles, and organizational guidelines for FID**

• **Adds paragraph on United States FID capabilities**

• **Introduces FID assessment paragraph**

• **Expands training chapter into a section on training joint forces for FID and a section on training host nation forces**

• **Provides discussion of site survey considerations and health service and medical civil-military operations as part of employment considerations**

• **Integrates psychological operations (PSYOP) paragraph into the direct support section, explaining PSYOP goals within FID**

• **Incorporates discussion of security force assistance (SFA) and explains the relationship between FID and SFA.**

• **Adds section on transition and redeployment, to include termination of operations, terminations approaches, military considerations, mission handoff procedures, and post mission debriefing procedures**

• **Removes appendices on civil affairs estimate of the situation, PSYOP and estimate of the situation, and health service support**

• **Updates the definitions of FID, security assistance, and security cooperation organization**

• **Creates new terms and definitions security force and security force assistance**

• **Removes the term and definition for security assistance organization**

Intentionally Blank

TABLE OF CONTENTS

EXECUTIVE SUMMARY
COMMANDER'S OVERVIEW

- **Provides an Overview of Foreign Internal Defense**

- **Discusses Host Nation Internal Defense and Development**

- **Outlines Organization and Responsibilities for Foreign Internal Defense**

- **Discusses Planning for Foreign Internal Defense**

- **Covers Training for Joint and Host Nation Forces**

- **Discusses Foreign Internal Defense Operations**

Overview of Foreign Internal Defense

Foreign internal defense (FID) refers to activities that support a host nation (HN) internal defense and development (IDAD) strategy.

Foreign internal defense (FID) is the participation by civilian and military agencies of a government in any of the action programs taken by another government or other designated organization, to free and protect its society from subversion, lawlessness, insurgency, terrorism, and other threats to their security. The focus of US FID efforts is to support the host nation's (HN's) internal defense and development (IDAD), which can be described as the full range of measures taken by a nation to promote its growth and protect itself from the security threats described above.

While historically, the United States provided notable, and largely unconditional, assistance to friendly foreign nations following World War II (Truman Doctrine and the Marshall Plan), FID as conducted today has its genesis in the post-Vietnam era when US policy shifted to emphasize that the United States would assist friendly nations, but would require them to provide the manpower and be ultimately responsible for their own national defense.

It is important to frame the US FID effort within the context of the US doctrine it supports and to understand how it fits into the HN IDAD program. US military support to FID should focus on assisting an HN in anticipating, precluding, and countering threats or potential threats and addressing the root causes of instability. Emphasis on internal developmental programs as well as internal defense programs when organizing, planning, and executing military support to US FID activities is essential. Although the FID operation is considered military engagement, security cooperation, and deterrence, FID may include or support operations from across the range of military operations to support the HN's IDAD strategy. Accordingly, US military operations supporting FID provide training, materiel, advice, or assistance to local forces executing an IDAD program, rather than US forces conducting IDAD military missions for the HN.

The FID effort is a multinational and interagency effort, requiring integration and synchronization of all instruments of national power.

Military officials often have greater access to, and credibility with, HN regimes that are heavily influenced or dominated by their own military. However, the characteristics of FID involve the instruments of national power beyond the military instrument (diplomatic, informational, and economic) through which sources of US power (such as financial, intelligence, and law enforcement) can be applied to support an HN IDAD program.

The Department of Defense (DOD) employs a number of FID tools. **Security cooperation** is DOD interactions with foreign defense establishments to build defense relationships that promote specific US security interests, develop allied and friendly military capabilities for self-defense and multinational operations, and provide US forces with peacetime and contingency access to a host nation. The *Guidance for Employment of the Force* contains DOD guidance for SC. This guidance provides goals and activities for specific regions and provides the overarching framework for many FID-related activities. **Indirect support,** employing security assistance (SA), military exchange programs, and joint and multinational exercises, focuses on building strong national infrastructures through economic and military capabilities that contribute to self-sufficiency. **Direct support (not involving combat operations)** involves the use of US forces normally focused on civil-military operations (CMO) (primarily the provision of services to the local populace), psychological operations (PSYOP), communications and intelligence cooperation, mobility, and logistic support. The final tool is **US combat operations** in support of FID operations, which requires a Presidential decision and serves only as a temporary solution until HN forces are able to stabilize the situation and provide security for the populace.

Internal Defense and Development

An IDAD program focuses on building viable political, economic, military, and social institutions that respond to the needs of the (HN's) society.

The construct of an IDAD program should integrate security force and civilian actions into a coherent, comprehensive effort. The strategy, therefore, includes measures to maintain conditions under which orderly development can take place. Furthermore, the successful IDAD strategist must realize that the true nature of the threat to the government lies in the adversary's political strength rather than military power. Military and paramilitary programs are necessary for success, but are not sufficient alone.

An IDAD program blends four interdependent functions to prevent or counter internal threats. **Balanced development** attempts to achieve national goals through political, social, and economic programs, allowing all individuals and groups in the society to share in the rewards of development, thus alleviating frustration. The **security** function includes all activities implemented in order to protect the populace from violence and to provide a safe environment for national development. The security effort should establish an environment in which the HN can provide for its own security with limited US support. **Neutralization,** a political concept, physically and psychologically separates an insurgent or criminal element from the population, thereby making threatening elements irrelevant to the political process. It includes all lawful measures (except those that degrade the government's legitimacy) to discredit, disrupt, preempt, disorganize, and defeat the insurgent organization. **Mobilization** provides organized manpower and materiel resources and includes all activities to motivate and organize popular support of the government. Mobilization allows the government to strengthen existing institutions, to develop new ones to respond to demands, and promotes the government's legitimacy.

Certain principles guide efforts in the four functional areas in the IDAD strategy to prevent or defeat an internal threat. The principles are **unity of effort**, requiring cooperation among all forces and agencies toward a commonly recognized objective; **maximum use of intelligence** requires that all operations be based on reliable, accurate, and timely intelligence to protect friendly FID operations and to counter and penetrate opposing force intelligence collection operations; **maximum use of CMO and PSYOP,** which helps generate active and tacit popular support of the HN government and buys the time for the HN civil authorities and government to eliminate or mitigate valid popular grievances; **minimum use of force**, appropriate (and proportional) to the incident at hand, to maintain order and incorporate economy of force; **a responsive government**, whose ability to mobilize manpower and resources as well as motivate the people reflects its administrative and management capabilities; and **use of strategic communication,** characterized by defense support of public diplomacy, public affairs (PA), and information operations (IO) messages that are coordinated early during the planning process and continually throughout the operation and that are derived from harmonized themes lest the credibility of some of the agencies involved in the FID effort, particularly the HN's, becomes compromised or lost.

The **organization** that is created to coordinate, plan, and conduct IDAD activities may vary from country to country in order to adapt to existing conditions. The organization should be structured and chartered so that it can coordinate and direct the IDAD efforts of existing government agencies; however, it should minimize interference with those agencies' normal functions. A **national-level** organization's major offices normally correspond to branches and agencies of the national government concerned with insurgency, illicit drug trafficking, and terrorist or other internal threats. At the **subnational level,** area coordination centers (ACCs) may function as multinational civil-military headquarters at subnational, state, and local levels. ACCs perform a twofold mission: they provide integrated planning, coordination, and direction for all internal defense efforts and they facilitate an immediate, coordinated response to operational requirements. There are two types of subnational ACCs that a government may form — regional and urban. **Regional ACCs** normally locate with the nation's first subnational political subdivision with a fully developed governmental apparatus (state, province, or other). Where higher level ACCs are lacking, **urban ACCs** that include representatives from local public service agencies, such as police, fire, medical, public works, public utilities, communications, and transportation, are appropriate for cities and heavily populated areas.

Organization and Responsibilities for Foreign Internal Defense

The lines of organization and coordination during FID operations are complex.

For FID to be successful in meeting an HN's needs, the United States government (USG) must integrate the efforts of multiple government agencies. Such integration and coordination are essentially vertical between levels of command and organization, and horizontal between USG agencies and HN military and civilian agencies. Management of the FID effort begins at the national level, with the selection of those nations the US will support through FID efforts. This decision is made by the President with advice from the Secretary of State, Secretary of Defense (SecDef), and other officials. The US will consider FID support when the existing or threatened internal disorder threatens US national strategic goals, or when the threatened nation requests and is capable of effectively using US assistance.

The **National Security Council** (NSC) will generally provide the initial guidance and translation of national-level decisions pertaining to FID. The **Department of State** (DOS) is generally the lead government agency and assists the NSC in building and carrying out national FID policies and priorities. The **United States Agency for International Development** carries out nonmilitary assistance programs designed to assist certain less developed nations to increase their productive capacities and improve their quality of life. The **Director of National Intelligence and the Director of the Central Intelligence Agency** support the FID mission in both a national-level advisory capacity and at the regional and country levels through direct support of FID activities.

The **Department of Defense** national-level organizations involved in FID management include the

Office of the Secretary of Defense (OSD) and the Joint Staff. OSD acts as a policy-making organization in most FID matters. The Under Secretary of Defense for Policy exercises overall direction, authority, and control concerning SA for OSD through the various assistant secretaries of defense. The Defense Security Cooperation Agency (DSCA) is the principal DOD organization through which SecDef carries out responsibilities for SA, conducting international logistics and sales negotiations and serving as the DOD focal point for liaison with US industry regarding SA. Finally, DSCA develops and promulgates SA procedures, maintains the database for the programs, and makes determinations with respect to the allocation of foreign military sales administrative funds.

The Chairman of the Joint Chiefs of Staff (CJCS) plays an important role in providing strategic guidance to the combatant commanders for the conduct of military operations to support FID. This guidance is provided primarily through the National Military Strategy (NMS) and the Joint Strategic Capabilities Plan (JSCP), the key components of the Joint Strategic Planning System (JSPS).

United States Coast Guard, within the Department of Homeland Security is specifically authorized to assist other federal agencies in the performance of any activity for which especially qualified, including SA activities for DOS and DOD.

Geographic combatant commanders (GCCs) are responsible for planning and executing military operations in support of FID within their area of responsibility (AOR). **Other combatant commanders** play a supporting role by providing resources to conduct operations as directed by the President or SecDef. All staff elements contribute to the overall support of the FID operation. For example, the **plans directorate** incorporates military support to FID into theater strategy and plans; the **operations directorate** monitors the execution of military operations in support of FID; and the **intelligence directorate** produces intelligence that often supplements estimates produced by the national intelligence agencies. **Other staff functions** may be given primary responsibility for specific military technical support missions and will usually focus on the direct support (not involving combat operations) category of military support to FID.

When authorized by SecDef through the CJCS, commanders of unified commands may establish **subordinate unified commands** (also called subunified commands), either geographic or functional, to conduct operations on a continuing basis in accordance with the criteria set forth for unified commands. Theater special operations commands are of particular importance because of the significant role of special operations forces (SOF) in FID operations. Coordination between the joint force special operations component commander and the other component commanders within the combatant command is essential for effective management of military operations in support of FID, including joint and multinational exercises, mobile training teams, integration of SOF with conventional forces, and other operations.

The President gives the chief of the **diplomatic mission,** normally an ambassador, full responsibility for the direction, coordination, and supervision of all USG executive branch employees in-country. Close coordination with each chief of mission (COM) and country team is essential in order to conduct effective, country-specific FID operations that support the HN's IDAD program and US regional goals and objectives. The principal military member of the **country team** is the senior defense official/defense attaché (SDO/DATT), who functions as the COM's principal military advisor on defense and national security issues, the senior diplomatically accredited DOD military officer assigned to a US diplomatic mission, and the single point of contact for all DOD matters involving the embassy or DOD elements assigned to or working from the embassy. Additionally, in the majority of countries, the functions of a security cooperation organization (SCO) are carried out under the direction of the SDO/DATT. The SCO is the most important FID-related military activity under the supervision of the ambassador. The specific title of the SCO may vary; however, these differences reflect nothing more than the political climate within the HN.

As examples, an SCO may be referred to as a military assistance advisory group, military advisory group, office of military cooperation, or office of defense cooperation.

Planning for Foreign Internal Defense

FID planning is designed to bolster the internal stability and security of the supported nation.

A comprehensive planning process at both the national and theater level is vital in order to provide the means to bolster the internal stability and security of the supported nation. The type of planning necessary is dictated by the type or types of support being provided. Support in anticipating and precluding threats is preventive in nature and is likely to require a mix of indirect support and direct support not involving combat operations. Depending on whether the mission has originated through DOD or DOS, how, where, and at what level the planning, coordination, and resourcing takes place will vary.

FID has certain aspects that make planning for it complex.

Some basic imperatives when integrating FID into strategies and plans are:

Maintain HN Sovereignty and Build Legitimacy. Ultimately, FID operations are only as successful as the HN's IDAD program.

Understand long-term or strategic implications and sustainability of all US assistance efforts. Building HN development and defense self-sufficiency may require large investments of time and materiel.

Tailor military support to FID for the operational environment and the specific needs of the supported HN. The tendency to use a US frame of reference can result in equipment, training, and infrastructure not at all suitable for the nation receiving US assistance.

Ensure Unity of Effort/Unity of Purpose. Planning should consider and, where appropriate, integrate *all* instruments of national power and intergovernmental organizations, nongovernmental organizations, and HN capabilities in order to reduce inefficiencies and enhance strategy in support of FID and HN IDAD efforts.

Understand US Foreign Policy. NSC directives, plans, or policies are the guiding documents; however, US policy may change as a result of developments in the HN or broader political changes in either country. DOD planners should seek guidance from the COM and country team in interpreting foreign policy and guiding US efforts in a particular country.

Understand the Information Environment. In an environment characterized by "instant communications," proactive PA and PSYOP programs can address regional, transregional, and even global audiences that may have (or perceive they have) a stake in US FID operations.

Sustain the Effort. Plan for the US sustainment effort as well as the efforts necessary for the HN to sustain its operations after the US or multinational forces depart.

GCCs base strategy and military planning to support FID on the broad guidance and missions provided in the JSPS. The NMS supports the aims of the National Security Strategy (NSS) and implements the national defense strategy. Through the guidance and resources provided in the JSCP, the GCCs develop their operation plan and operation plan in concept format to support FID. The *Guidance for Employment of the Force* provides the foundation for all DOD interactions with foreign defense establishments, and supports the President's NSS. The GCC, using an integrated priority list, also identifies requirements to support FID efforts and request authorization and resourcing.

Military activities in support of FID requirements are integrated into concepts and plans from the strategic level down to the tactical level. Theater strategy translates national and alliance strategic tasks and direction into long-term, regionally focused operational tasks and direction to accomplish specific missions and objectives. In peacetime, FID is an integral part of the strategy of deterring hostilities and enhancing stability in the theater. The GCC's theater campaign plan is the primary document that focuses on the command's steady-state activities, which include operations, SC, and other activities designed to achieve theater strategic end states.

Training

The strategic imperative to assist allies, coalition partners, and the governments of threatened states in resisting aggression demands that joint forces strengthen their abilities to assist, train, and advise foreign military and security forces.

Joint forces training requirements and skills needed for successful military operations in support of FID include: overall US and theater goals for FID; area and cultural orientation; language training; standards of conduct; relationships of FID programs to intelligence collection; coordinating relationships with other USG agencies; legal guidelines; rules of engagement (ROE); and tactical force protection training. Training to prepare for military operations to support FID requires that a broad range of areas be covered. The training also must be designed to support a mix of personnel, ranging from language-trained and culturally focused SOF to those totally untrained in the specific area where the FID program is located. A combination of institutional and unit-conducted individual and collective training will be required.

In addition to training for the joint force preparing to conduct FID, FID operations include training of HN security forces to build the capacity to support the IDAD strategy. After completing a thorough assessment and site survey, the JFC develops a training plan based on a thorough mission analysis and assessment of the IDAD strategy, HN capabilities and needs, and the operational environment. This plan should be developed in conjunction with both the country team and with commanders of HN forces to ensure that comprehensive objectives are detailed.

Foreign Internal Defense Operations

FID operational activities emphasize interagency coordination.

Several areas deserve special attention when discussing employment of forces in FID operations, including IO impact, psychological impact, intelligence support, SOF and/or conventional force selection, public information programs, logistic support, counterdrug operations in FID, combating weapons of mass destruction operations in FID, counterterrorism and FID, operations security, and lessons learned.

Additionally, units assigned a FID mission must implement procedures to help DOS and the country team vet HN forces to ensure the identification of personnel with a history of human rights violations.

The military will primarily provide equipment, training, and services to the supported HN forces.

Indirect Support. GCCs are active in the SA process by advising ambassadors through the security assistance organization (SAO) and by coordinating and monitoring ongoing SA efforts in their AORs. In addition, through coordination with HN military forces and supporting SAOs, the combatant commander can assist in building credible military assistance packages that best support long-term goals and objectives of regional FID programs. **Joint and multinational exercises** can enhance a FID program. They offer the advantage of training US forces while simultaneously increasing interoperability with HN forces and offering limited HN training opportunities. **Exchange programs** foster greater mutual understanding and familiarize each force with the operations of the other. Exchange programs are another building block that can help a commander round out his FID plan. Some of these programs include reciprocal unit exchange programs, personnel exchange programs, individual exchange programs, and combination programs.

Direct support involves US forces actually conducting operations in support of the HN.

Direct Support Not Involving Combat Operations. Direct support operations provide immediate assistance and are usually combined in a total FID program with indirect operations. Several types of direct operations are important to supporting FID. **CMO** span a very broad area in FID and include activities such as civil affairs activities, foreign humanitarian assistance, humanitarian and civic assistance, and military civic action across the range of military operations. Using CMO to support military activities in a FID program can enhance preventive measures, reconstruction efforts, and combat operations in support of an HN IDAD program. **PSYOP** supports the achievement of national objectives by influencing behaviors in select target foreign audiences. **Military training to HN forces may focus on** subversion, lawlessness, and insurgency problems encountered by the HN that may be beyond its capabilities to control. **Logistic support** operations are limited by US law and usually consist of transportation or limited maintenance support, although an acquisition and cross-servicing agreement can allow additional support in areas beyond those. Authorization for combatant commanders to provide logistic support to the HN military must be received from the President or SecDef. **Intelligence and information sharing,** although two separate areas, are closely related and have many of the same employment considerations. Assistance may be provided in terms of evaluation, training, limited information exchange, and equipment support.

US participation in combat operations as part of a FID effort requires Presidential authority.

Combat Operations. Many considerations, including CMO and PSYOP, must be discussed and reviewed when employing combat forces in support of FID. They include: maintaining close coordination with the HN IDAD organization; tiering of forces; establishing transition points; maintaining a joint, interagency, and multinational focus; identifying and integrating logistics, intelligence, and other combat support means in US combat operations; conduct combat operations only when directed by legal authority to stabilize the situation and to give the local government and HN military forces time to regain the initiative; strict adherence to respect for human rights; following the ROE; preventing indiscriminate use of force; maintaining the US joint intelligence network; and integrating with other FID programs. The command and control relationships will be modified based on the political, social, and military environment of the area. The HN government and security forces must remain in the forefront. Finally, sustainment of US forces is essential to success. Political sensitivities and concern for HN legitimacy and minimum US presence do change the complexion of sustainment operations in FID.

Transition and Redeployment. Redeployment of units conducting FID operations does not typically indicate the end of all FID operations in the HN. Rather, in long-term FID operations as security and other conditions improve and internal threats become manageable for HN personnel, direct military-to-military activities by units will continue, but these activities may become more intermittent with gaps between regular exercises and exchanges. In both immediate mission handoff and intermittent FID operations, capturing lessons learned in thorough postmission debriefings is essential to continue to build institutional FID knowledge and refine FID doctrine and training.

CONCLUSION

This publication establishes joint doctrine for the Armed Forces of the United States involved in or supporting FID. It discusses how joint operations, involving the application of all instruments of national power, support HN efforts to build capability and capacity to free and protect its society from subversion, lawlessness, and insurgency.

Intentionally Blank

CHAPTER I
INTRODUCTION

"Although on the surface, FID [foreign internal defense] appears to be a relatively simple concept, that appearance is deceptive; FID is a much more nuanced and complicated operation than its definition at first implies. FID is often confused with or equated to training foreign forces, when in reality, there is much more to it."

Lieutenant Colonel John Mulbury
ARSOF [Army Special Operations Forces],
General [Conventional] Purpose Forces and FID
Special Warfare, January-February 2008

1. General

a. Military engagement, security cooperation (SC), and deterrence encompass a wide range of actions where the military instrument of national power supports other instruments of national power to protect and enhance national security interests and deter conflict. Within this range of military operations, **nation assistance (NA)** is civil or military assistance (other than foreign humanitarian assistance [FHA]) rendered to a nation by US forces within that nation's territory during peacetime, crises or emergencies, or war, based on agreements mutually concluded between the United States and that nation. NA operations support the host nation (HN) by promoting sustainable development and growth of responsive institutions. The goal is to promote long-term regional stability. **NA programs include security assistance (SA), humanitarian and civic assistance (HCA), and foreign internal defense (FID). FID is the participation by civilian and military agencies of a government in any of the action programs taken by another government or other designated organization, to free and protect its society from subversion, lawlessness, insurgency, terrorism, and other threats to their security.** Internal threats in the context of this publication means threats manifested within the internationally recognized boundaries of a nation. These threats can come from, but are not limited to, subversion, insurgency (including support to insurgency), and/or criminal activities.

b. The focus of US FID efforts is to support the HN's internal defense and development (IDAD). IDAD is the full range of measures taken by a nation to promote its growth and protect itself from subversion, lawlessness, insurgency, terrorism, and other threats to their security. It focuses on building viable institutions that respond to the needs of society. It is important to understand that both FID and IDAD, although defined terms and used throughout this publication, are not terms used universally outside the Department of Defense (DOD). Other terms could be used to encompass what are called FID and IDAD herein.

c. Military engagement during FID supports the other instruments of national power through a variety of activities across the range of military operations. In some cases, direct military support may be necessary in order to provide the secure environment for

IDAD efforts to become effective. However, absent direction from the President or the Secretary of Defense (SecDef), US forces engaged in NA are prohibited from engaging in combat operations, except in self-defense.

d. From the US perspective, **FID refers to the US activities that support an HN IDAD strategy designed to protect against subversion, lawlessness, insurgency, terrorism, and other threats to their security, consistent with US national security objectives and policies.**

2. Background

a. The United States has a long history of assisting the governments of friendly nations facing internal threats. In the chaos after World War II, the United States followed its massive wartime Lend Lease Program with postwar assistance through the Marshall Plan by providing up to 90 percent of the support to the United Nations Relief and Rehabilitation Administration in support of war-stricken Europe and the Far East. New balances of power and the devastation of Europe had permanently changed the strategic role and interests of the United States. This new role was vividly demonstrated by the economic, equipment, training, and advisory support provided to Greece and Turkey to stabilize their governments. This postwar US policy was reflected in the staunchly anticommunist Truman Doctrine that established US policy as that of "…supporting free peoples who are attempting to resist subjugation by armed minorities or by outside pressures."

b. Both the Truman Doctrine and Marshall Plan were US-designed and US-implemented programs concentrating on repelling the external threat of communist aggression as well as thwarting internal threats to supported nations. Although this concept of support differs significantly from today's FID concept, these programs set the precedent for US support and assistance to friendly nations facing threats to their national security. Early postwar arms transfers were carried out as grant aid (giveaway) under the Military Assistance Program, often through a military assistance advisory group. Later, as the economies of recipient nations regenerated, arms transfers, economic aid, and collective security began to merge under a program that was to be known under the Nixon administration as SA.

c. The US policy of assisting friendly nations to develop stable governments and prevent the spread of communism continued through the US experience with Cuba beginning in 1959. This policy reached a new level during the 1960s with the Alliance for Progress, which involved Central and South America. This era of assistance culminated with the war in Vietnam, a major turning point for US policy that shaped the concept of FID as the United States now conducts it. In 1969, with US public and congressional opinion moving strongly against the war in Vietnam and against US intervention in general, President Richard M. Nixon announced a new US approach to supporting friendly nations. The Nixon Doctrine (also called the Guam Doctrine) expressed this policy with emphasis on the point that the US would assist friendly nations, but would require them to provide the manpower and be ultimately responsible for their own national defense. This principle of nations developing their own IDAD programs, supported through US training and materiel assistance, has become the basis for today's FID doctrine.

HISTORICAL CONTEXT OF FOREIGN INTERNAL DEFENSE

The concept of foreign internal defense as a way to provide US support to a host nation's internal defense and development (IDAD) plan originated from the Kennedy administration with National Security Action Memorandum (NSAM) 182 in August 1962. It was referred to as the "US Overseas Internal Defense Plan." This NSAM also provided the first US counterinsurgency (COIN) doctrine issued from the executive level. Resident within the narrative of this NSAM and COIN doctrine is the concept of the IDAD plan as central to aligning US support to other nations while striving to ensure that the strategic emphasis remains with the host nation government. In NSAM 341, the Johnson administration made minor adjustments to the approach, mainly structural in nature, as to how the National Security Council was organized, along with other minor changes, and re-issued NSAM 182 with the short title, "Foreign Internal Defense Policy," issued as executive policy, and as such, applicable to all government departments and agencies.

Various Sources

---

d. In recent times, the United States has provided the same type of NA in El Salvador, Colombia, Kuwait, the Philippines, Afghanistan, and the Republic of Georgia. Although not of the same magnitude as the post-World War II efforts, the United States contributed to restoring stability in the various regions after crisis situations.

e. Although FID is a core task of US Special Operations Command (USSOCOM) and special operations forces (SOF) maintain the capability to conduct such operations, conventional forces (CF) also possess capabilities to conduct FID. FID is not a military-only operation; rather, it includes an interagency approach to assisting an HN. The joint force commander (JFC) supporting a FID effort may employ capabilities provided by both CF and SOF. A robust FID operation may be conducted through the command and control (C2) structure of a joint task force (JTF) or a joint special operations task force (JSOTF). When CF and SOF are integrated, appropriate C2 or liaison elements should be exchanged or provided to the appropriate components of a joint force.

For additional detail on CF and SOF relationships, refer to Joint Publication (JP) 3-05.1, Joint Special Operations Task Force Operations, *and USSOCOM Pub 3-33,* Conventional Forces and Special Operations Forces Integration and Interoperability Handbook and Checklist, Version 2, September 2006.

3. Relationship of Foreign Internal Defense to Internal Defense and Development

a. It is important to frame the US FID effort within the context of the US doctrine it supports and to understand how it fits into the HN IDAD program.

b. As previously noted, FID falls under the NA construct. NA supports the HN by promoting sustainable development and growth of responsive institutions. The US goal is to promote long-term regional stability.

c. US military support to FID should focus on assisting an HN in anticipating, precluding, and countering threats or potential threats and addressing the root causes of instability. Emphasis on internal developmental programs as well as internal defense programs when organizing, planning, and executing military support to US FID activities is essential.

d. **US military involvement in FID has traditionally been focused toward counterinsurgency (COIN).** Although much of the FID effort remains focused on this important area, US activities may aim at other threats to an HN's internal stability, such as civil disorder, illicit drug trafficking, and terrorism. These threats may, in fact, predominate in the future as traditional power centers shift, suppressed cultural and ethnic rivalries surface, and the economic incentives of illegal drug trafficking continue. **Focusing on the internal development portion of IDAD expands the focus beyond COIN.**

For additional information on COIN, refer to JP 3-24, Counterinsurgency Operations.

e. US military operations supporting FID provide training, materiel, advice, or assistance to local forces executing an IDAD program, rather than US forces conducting IDAD military missions for the HN. Military operations are, at least to some degree, intertwined with foreign assistance provided by non-DOD agencies in the form of development assistance, humanitarian assistance, or SA described in legislation such as the Foreign Assistance Act (FAA).

4. **Foreign Internal Defense Within the Range of Military Operations**

a. Use of joint capabilities in military engagement, SC, and deterrence activities helps shape the operational environment and keep the day-to-day tensions between nations or groups below the threshold of armed conflict while maintaining US global influence. NA operations, including FID, are applied to meet military engagement, SC, and deterrence objectives.

b. FID operations occur throughout the range of military operations, most often as persistent and enduring cooperative security engagements, including SA and other shaping activities, but can function during limited contingency operations and larger scale activities to support an HN's COIN operations. Although the FID operation is considered military engagement, SC, and deterrence, FID may include or support operations from across the range of military operations to support the HN's IDAD strategy. For instance, FID in one country may include training the HN to improve functions to deal with criminal activity such as drug trafficking, while in another country FID may include training the HN to conduct COIN operations. Although FID may include, as part of supporting the HN's IDAD strategy, capacity building across the HN, the focus is on combating internal threats to assist the HN in maintaining legitimacy and influence over the relevant population, especially as it pertains to irregular warfare (IW). IW is a primary focus of FID.

For additional detail on the range of military operations and/or IW, refer to JP 3-0, Joint Operations.

5. **The Foreign Internal Defense Operational Framework and Instruments of National Power**

a. As shown in Figure I-1, characteristics of FID involve the instruments of national power (diplomatic, informational, military, and economic) through which sources of US power (such as financial, intelligence, and law enforcement) can be applied to support an HN IDAD program. Although this publication centers on the military instrument's contribution, it is also important to understand the overlying national strategy that directs FID activities and how all instruments of national power support these activities.

b. The FID operational framework, shown in Figure I-2, is discussed in more detail throughout the remainder of this chapter.

c. **Diplomatic Instrument.** FID makes extensive use of the diplomatic instrument of national power. A dysfunctional political system in a nation results in internal instability. Diplomacy is often the first instrument exercised by the United States and depending upon the decisions made by the President or the President's designee, the

CHARACTERISTICS OF FOREIGN INTERNAL DEFENSE

✓ **Involves all instruments of national power**

✓ **Can occur across the range of military operations**

✓ **Is conducted by both conventional forces and special operations forces**

✓ **Supports and influences the host nation's internal defense and development program**

✓ **Includes training, materiel, technical and organizational assistance, advice, infrastructure development, and tactical operations**

✓ **Generally, the preferred methods of support are through assistance and development programs**

Figure I-1. Characteristics of Foreign Internal Defense

THE FOREIGN INTERNAL DEFENSE FRAMEWORK, INSTRUMENTS OF NATIONAL POWER, AND SELECTED SOURCES OF POWER

DIPLOMATIC

Foreign internal defense (FID) makes extensive use of the diplomatic instrument of national power and is often the first instrument exercised by the United States.

INFORMATIONAL

Effective use of public diplomacy, public affairs activities, and psychological operations is essential to FID. Accurate portrayal of United States FID efforts through positive information programs can influence worldwide perceptions of the FID efforts and the host nation's (HN's) desire to embrace changes and improvements necessary to correct its problems.

MILITARY

The military plays an important supporting role in FID. Military FID activities can generally be categorized into:

- Indirect Support: FID operations that emphasize building strong national infrastructures through economic and military capabilities that contribute to self-sufficiency.
- Direct Support (not involving combat operations): FID operations providing direct assistance to the HN civilian populace or military when the HN has not attained self-sufficiency and is faced with threats beyond its capability to handle.
- Combat Operations.

ECONOMIC

Economics influence every aspect of FID support. It is used in a variety of ways, ranging from direct financial assistance and favorable trade arrangements, to the provision of foreign military financing under security assistance.

Selected Sources of Power Applied Through the Instruments

FINANCIAL

This involves United States Government (USG) agencies working with the governments of other nations and international financial institutions to encourage economic growth; raise standards of living; and to the extent possible, predict, prevent, or limit economic and financial crises.

INTELLIGENCE

This seeks to provide national leadership with information to help achieve national goals and objectives and to provide military leadership with information to accomplish missions and implement national security strategy. Attention is focused to identify adversary capabilities and centers of gravity; protect friendly course of action; and to assist planning of friendly force employment. Whereas informational power projects information to shape environments, intelligence seeks to gather information to understand environments and to inform decisionmaking.

LAW ENFORCEMENT

The USG is accountable to its people and expected to govern effectively through administration and enforcement of the law. This also requires ensuring public safety against foreign and internal threats; preventing and controlling crime; punishing unlawful behavior; and fair and impartial administration of justice. Because the threats to US security and public safety are global, extensive work is required outside US borders to combat and counter these threats.

Figure I-2. The Foreign Internal Defense Framework, Instruments of National Power, and Selected Sources of Power

diplomatic instrument of national power may be the only practical instrument of national power that can be brought to bear. Indirect and direct military support provided through training, logistics, or other support all make significant diplomatic statements by demonstrating US commitment and resolve.

d. **Informational Instrument.** The United States Government (USG) uses strategic communication to provide top-down guidance relative to using the informational instrument of national power in specific situations. The predominant military activities that support USG strategic communication themes and messages—information operations (IO), public affairs (PA), and defense support to public diplomacy (DSPD)— are essential to FID. FID activities may lead to public misunderstanding and may be exploited by elements hostile to the United States and its allies. US foreign assistance (development assistance, humanitarian assistance, and SA) has often been met with skepticism by the American public and typically has been the target for adversary propaganda. **FID offers a mechanism to portray US support in a positive light, but not at the expense of the supported nation that may be sensitive to accepting aid.** Accurate portrayal of US FID efforts through positive information programs can influence worldwide perceptions of the US FID efforts and HN desire to embrace changes and improvements necessary to correct its problems as well as deter those opposed to such changes. Additionally, some HNs may be willing to accept informational or intelligence support more readily than more transparent military support to internal and external audiences.

e. **Military Instrument.** The military plays an important supporting role in the overall conduct of FID activities and this role cannot be conducted in isolation. In some cases, the role of the US military may become more important because military officials have greater access to, and credibility with, HN regimes that are heavily influenced or dominated by their own military. The ability of the US military to influence the professionalism of the HN military, and thus its democratic process, is considerable. In such cases, success may depend on US representatives being able to persuade host military authorities to lead or support reform efforts aimed at eliminating or reducing corruption and human rights abuse. **The FID effort is a multinational and interagency effort, requiring integration and synchronization of all instruments of national power.** US military support also requires joint planning and execution to ensure that the efforts of all participating combatant commands, subordinate joint force commands, and/or Service or functional components are mutually supportive and focused. **FID is conducted by both CF and SOF as appropriate. Generally, joint force activities that support FID are categorized as indirect support, direct support (not including combat operations), and combat operations.** See Figure I-3 for examples of indirect support, direct support, and combat operations. These categories represent significantly different levels of US diplomatic and military commitment and risk. It should be noted, however, that various activities and operations within these categories can occur simultaneously. As an example, certain forms of indirect support and direct support (not involving combat operations) may continue even when US forces are committed to a combat role.

Figure I-3. Foreign Internal Defense: Integrated Security Cooperation Activities

f. **Economic Instrument.** Economics influence every aspect of FID support. Often, the internal strife a supported nation faces is brought on by unfavorable economic conditions within that nation. These conditions weaken national infrastructures and contribute to instability, particularly when the HN government is perceived as not being able to meet the basic needs of the people. These instabilities may produce an environment ripe for increasing subversion, lawlessness, and insurgency. The economic tool is used in a variety of ways, ranging from direct financial assistance and favorable trade arrangements to the provision of foreign military financing (FMF) under SA.

g. **Selected Sources of Power Applied Through the Instruments of National Power**

(1) **Financial.** The financial activities of the nation promote the conditions for prosperity and stability in the United States and encourage prosperity and stability in the rest of the world. The Department of the Treasury (TREAS) is the primary federal

agency responsible for the economic and financial prosperity and security of the United States. In the international arena, TREAS works with other federal agencies, governments of other nations, and international financial institutions to encourage economic growth; raise standards of living; and predict and prevent, to the extent possible, economic and financial crises.

(2) **Intelligence.** Intelligence activities provide the national leadership with the information needed to realize national goals and objectives and to implement the National Security Strategy (NSS). It also provides the military leadership with the information needed to accomplish missions and implement the National Military Strategy (NMS). Planners use intelligence to identify the adversary's capabilities and centers of gravity, project probable courses of action (COAs), and assist in planning friendly force employment. Intelligence also provides assessments that help the JFC decide which forces to deploy; when, how, and where to deploy them; and how to employ them in a manner that accomplishes the mission at the lowest human and political cost. Intelligence support to USG FID may involve the identification of critical infrastructure, mapping of significant activities that may be threats, and products that support safety of navigation or provide increased situational awareness for strategic decisionmaking.

(3) **Law Enforcement**

(a) The mission of the Department of Justice (DOJ) is to enforce the law and defend the interests of the United States according to the law, to ensure public safety against both foreign and domestic threats, to provide federal leadership in preventing and controlling crime, to seek just punishment for those guilty of unlawful behavior, and to ensure fair and impartial administration of justice for all Americans. The Attorney General represents the United States in legal matters and advises the President and heads of the executive departments of the government when so requested. DOJ is the central agency for enforcement of federal laws.

(b) Terrorism poses a grave threat to individuals' lives and national security around the world. The International Criminal Police Organization (INTERPOL) (the largest international police organization) has therefore made available various training opportunities and resources to support member countries in their efforts to protect their citizens from terrorism, including bioterrorism, firearms and explosives, attacks against civil aviation, maritime piracy, and weapons of mass destruction (WMD). The United States National Central Bureau of INTERPOL operates in conjunction with the Department of Homeland Security and within the guidelines prescribed by DOJ.

(c) Other law enforcement agencies include the United States Secret Service, Drug Enforcement Administration (DEA), and the Federal Bureau of Investigation. Additionally, the Department of State (DOS) Bureau of International Narcotics and Law Enforcement Affairs (INL), while not a law enforcement agency, can be instrumental in the development of policies and programs to combat international narcotics and crime. The DOJ Bureau of Alcohol, Tobacco, and Firearms can also provide information about international crime and explosives trafficking.

(d) The United States Coast Guard (USCG) is a unique, multimission organization that is at all times a law enforcement agency (Title 14, United States Code [USC], Section 89) as well as a military service and a branch of the Armed Forces of the United States (Title 10, USC, Section 101; Title 14, USC, Sections 1 and 2). The ability to handle evolving scenarios as a federal law enforcement agency or an armed force is a unique characteristic of the USCG. Congress established the USCG as a maritime constabulary force of the United States by specifically assigning it the duty to enforce or assist the enforcement of all applicable federal laws on, under, and over the high seas and waters subject to the jurisdiction of the United States.

6. **Department of Defense Foreign Internal Defense Tools**

a. **Security Cooperation. SC involves all DOD interactions with foreign defense establishments to build defense relationships** that promote specific US security interests, develop allied and friendly military capabilities for self-defense and multinational operations, and provide US forces with peacetime and contingency access to an HN. SC is the means by which DOD encourages and enables countries and organizations to work with the United States to achieve strategic objectives. The *Guidance for Employment of the Force* (GEF) contains DOD guidance for SC. This guidance provides goals and activities for specific regions and provides the overarching framework for many FID-related activities. The GEF includes the SC tools/resources shown in Figure I-4, some of which will be discussed (as they apply to FID) further below. Additionally, Figure I-3 shows how SC encompasses FID, which consists of indirect support, direct support (not involving combat operations), and combat operations. Each of these is discussed below. Relationships among SC, SA, and FID are depicted at Figure I-5.

For further guidance on SC tools/resources, refer to the current GEF.

b. **Indirect Support.** These are FID operations that emphasize the principle of HN self-sufficiency. **Indirect support focuses on building strong national infrastructures through economic and military capabilities that contribute to self-sufficiency.** The US military contribution to this type of support is derived from SC guidance and provided primarily through SA, supplemented by multinational exercises, exchange programs, and selected joint exercises.

(1) **Security Assistance.** SA is a principal element in the US FID effort. Like FID itself, SA is a broad, encompassing topic and includes efforts of civilian agencies as well as those of the military. **SA is the provision of defense articles, military training, and other defense-related services by grant, loan, credit, or cash sales in furtherance of US national policies and objectives.** SA, while integral to FID, is much broader than FID alone. SA is predominately aimed at enhancing regional stability of areas of the world facing external vice internal threats. **Note that only a portion of the overall SA effort fits into the FID area, but that it is a large part of the overall FID indirect support effort.** Also, it is important to note that the direct support (not involving combat operations) category typically makes up the preponderance of the remaining military operations. The SA program is authorized by the FAA of 1961 as amended, and the

SECURITY COOPERATION ACTIVITIES

- Multinational Education
- Multinational Exercises
- Multinational Experimentation
- Multinational Training
- Counternarcotics Assistance
- Counter / Nonproliferation
- Defense and Military Contacts
- Defense Support to Public Diplomacy
- Facilities and Infrastructure Projects
- Humanitarian Assistance
- Intelligence Cooperation
- Information Sharing
- International Armaments Cooperation
- Security Assistance
- Other Programs and Activities

Figure I-4. Security Cooperation Activities

Arms Export Control Act (AECA) of 1976 as amended, and is under the supervision and general direction of DOS. **SA is the military component of foreign assistance implemented by DOD in accordance with policies established by DOS, and has as its principal components foreign military sales (FMS), FMF, international military education and training (IMET), peace operations (PO), and excess defense articles (EDA).** DOS provides financial support to international peacekeeping operations (PKO), a subset of PO, through a PKO fund. These components, combined with the Economic Support Fund and commercial sales licensed under the AECA, are SA tools that the United States can use to further its national interests and support the overall FID effort.

(a) **Foreign Military Sales.** FMS is a nonappropriated program through which foreign governments can purchase defense articles, services, and training from the United States. Eligible nations can use this program to help build national security infrastructures. A limitation of this program is that the nations that require assistance are often unable to finance their needs.

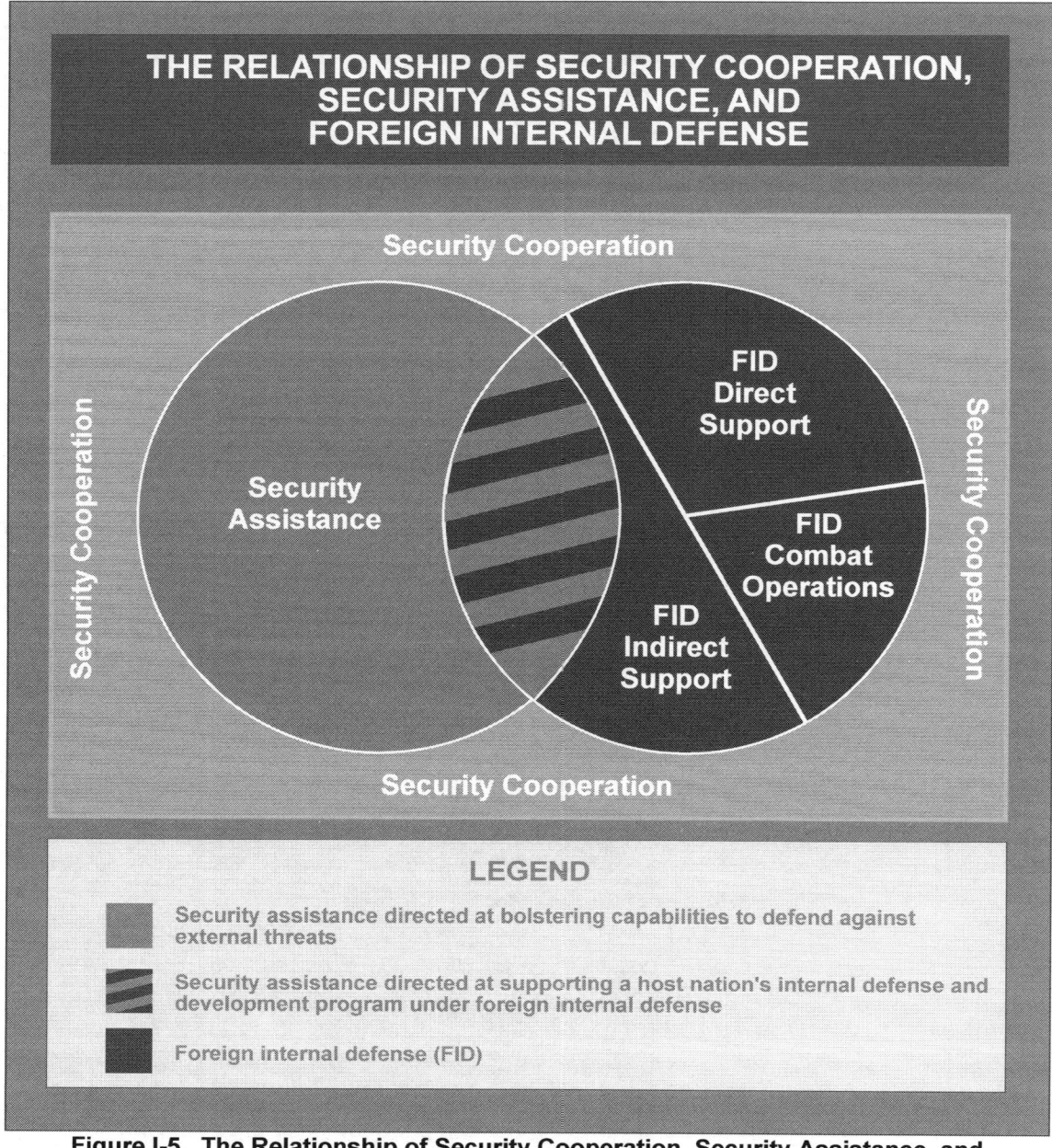

Figure I-5. The Relationship of Security Cooperation, Security Assistance, and Foreign Internal Defense

(b) **Foreign Military Financing.** FMF falls inside the military assistance budget process of DOS. FMF provides funding to purchase defense articles and services, design and construction services, and training through FMS or commercial channels. FMF can be an extremely effective FID tool, providing assistance to nations with weak economies that would otherwise be unable to afford US assistance.

(c) **International Military Education and Training.** IMET, also part of the DOS military assistance budget process, contributes to internal and external security of a country by providing training to selected foreign militaries and related civilian personnel on a grant aid basis. The program helps to strengthen foreign militaries through US military training (and exposure to culture and values) that are necessary for the proper functioning of a civilian controlled, apolitical, and professional military. This program has long-term positive effects on US and HN bilateral relations. IMET serves as an influential foreign policy tool where the US shapes doctrine, promotes self-sufficiency in maintaining and operating US acquired defense equipment, encourages the value of rule of law, and occasionally has a marked effect on the policies of the recipient governments. Foreign students—many of whom occupy the middle and upper echelons of their country's military and political establishments—are taught US defense doctrine and employment of US weapon systems, values, and management skills, resulting in greater cooperation and interoperability.

(d) **Peace Operations.** This program funds US PO, such as the multinational force and observers in the Sinai and the US contribution to the United Nations Force in Cyprus. These operations are limited in scope and funding levels and, although related to FID operations, are generally considered separate activities with very focused goals and objectives.

For further information on PO, refer to JP 3-07.3, Peace Operations.

(e) **Excess Defense Articles.** EDA are DOD and USCG-owned defense articles no longer needed and declared excess by the Armed Forces of the United States. This excess equipment is offered at reduced or no cost to eligible foreign recipients on an "as is, where is" basis. EDA are used for various purposes including internal security (including antiterrorism [AT] and nonproliferation). The EDA program works best in assisting friends and allies to augment current inventories of like items with a support structure already in place.

For further information on SA, refer to DOD 5105.38-M, Security Assistance Management Manual.

> *Security assistance has been and still remains an important instrument of US foreign policy. Arms transfers and related services have reached enormous dimensions and involve most of the world's nations, either as a seller/provider or buyer/recipient.*
>
> **Secretary of State Colin Powell,**
> **Written testimony for the Senate Foreign Relations Committee, 8 March 2001**

(2) **Exchange Programs.** Military exchange programs also support the overall FID effort by fostering mutual understanding between forces; familiarizing each force with the organization, administration, and operations of the other; and enhancing cultural awareness. Exchange programs, coupled with the IMET program, are extremely valuable in improving HN and US relations and may also have long-term implications for strengthening democratic ideals and respect for human rights among supported governments. It is important, however, that such exchange programs (funded with Title

10, USC, monies) do not themselves become vehicles for SA training or other services to the HN in contravention of the FAA and AECA.

(3) **Joint and Multinational Exercises.** These programs, which fall under the geographic combatant commander's (GCC's) theater campaign plan (TCP), strengthen US-HN relations and interoperability of forces. They are joint- and Service-funded and complement SA and civil-military operations (CMO) by validating HN needs and capabilities and by providing a vehicle for the conduct of HCA programs. There are very strict legal restrictions on the type of support that can be provided and on the monetary limits of such support.

For further information on CMO support to FID, refer to Chapter VI, "Foreign Internal Defense Operations," paragraph 10, "Civil-Military Operations;" Appendix A, "Legal Considerations;" and JP 3-57, Civil-Military Operations.

c. **Direct Support (Not Involving Combat Operations).** **These operations involve the use of US forces providing direct assistance to the HN civilian populace or military.** They differ from SA in that they are joint- or Service-funded, do not usually involve the transfer of arms and equipment, and do not usually but may include training local military forces. Direct support operations are normally conducted when the HN has not attained self-sufficiency and is faced with social, economic, or military threats beyond its capability to handle. **Assistance will normally focus on CMO (primarily, the provision of services to the local populace), psychological operations (PSYOP), communications and intelligence cooperation, mobility, and logistic support.** In some cases, training of the military and the provision of new equipment may be authorized.

(1) **Civil-Military Operations.** The purpose of CMO is to facilitate military operations, and to consolidate and achieve operational US objectives, through the integration of civil and military actions while conducting support to civil administration, populace and resources control, FHA, NA, and civil information management. CMO are the activities of a commander that establish, maintain, influence, or exploit relations among military forces, governmental and nongovernmental civilian organizations and authorities, and the civilian populace in a friendly, neutral, or hostile operational area in order to facilitate military operations, and to consolidate and achieve operational US objectives. CMO not only can support a COIN program, but also can enhance all FID activities. CMO may be used in a preventive or rehabilitative manner in order to address root causes of instability; in a reconstructive manner after conflict; or in support of disaster relief, consequence management, civil defense, counterdrug (CD), and AT activities.

(a) **Civil Affairs Operations (CAO).** CAO are those military operations conducted by civil affairs (CA) forces that enhance the relationship between military forces and civil authorities in localities where military forces are present; require coordination with other interagency organizations, intergovernmental organizations (IGOs), nongovernmental organizations (NGOs), indigenous populations and institutions, and the private sector; and involve application of functional specialty skills that normally are the responsibility of civil government to enhance the conduct of CMO. In FID, CAO facilitates the integration of US military support into the overall IDAD programs of the supported nation.

For further information on CMO support to FID, refer to Chapter VI, "Foreign Internal Defense Operations," *paragraph 10,* "Civil-Military Operations," *and to JP 3-57,* Civil-Military Operations.

(b) **Foreign Humanitarian Assistance.** FHA operations relieve or reduce the impact of natural or man-made disasters or other endemic conditions such as human suffering, disease, or privation in countries or regions outside the United States. FHA provided by US forces is limited in scope and duration. Although not part of NA, FHA may complement FID.

For further information on FHA, refer to JP 3-29, Foreign Humanitarian Assistance.

(c) **Humanitarian and Civic Assistance.** HCA is assistance to the local populace provided in conjunction with authorized military operations as specifically authorized by Title 10, USC, Section 401. Assistance provided under these provisions must promote the security interests of both the United States and the HN and the specific operational readiness skills of the members of the armed forces who participate in the activities. These activities must also be in accordance with HN laws. GCCs integrate and coordinate HCA activities within their areas of responsibility in the SC portion of their overall TCP. In the context of FID, HCA may be an important part of helping to build relationships between the local population and the joint force. When possible, HCA should give way to military civic action (MCA) as the FID operation progresses. In contrast to emergency relief conducted under FHA, HCA generally encompasses planned activities and includes:

<u>1.</u> **Medical, dental, and veterinary care** provided in rural or underserved areas of a country, including education, training, and technical assistance related to the care provided.

<u>2.</u> Construction of rudimentary **surface transportation systems.**

<u>3.</u> **Well drilling** and construction of basic **sanitation facilities.**

<u>4.</u> Rudimentary construction and repair of **public facilities** such as schools, health and welfare clinics, and other nongovernmental buildings.

<u>5.</u> Activities relating to the furnishing of education, training, and technical assistance concerning detection and clearance of landmines. Note: US forces are not to engage in the physical detection, lifting, or destroying of landmines (unless it is part of a concurrent military operation other than HCA).

(d) **Military Civic Action.** MCA is the use of predominantly indigenous military personnel to conduct construction projects, support missions, and services useful to the local population. These activities may involve US supervision and advice but will normally be conducted by the local military. MCA is an essential part of military support to FID to assist the local government in developing capabilities to provide for the security and well-being of its own population.

(2) **Psychological Operations.** The focus of joint military PSYOP objectives during FID operations is to **support US national objectives, to support the GCC's regional security strategy objectives, and to support the objectives of the country team.** Additionally, PSYOP is used to promote the ability of the HN to defend itself against internally and externally based insurgencies and terrorism by encouraging the civilian populace to actively support the HN military and government. PSYOP also may be used to modify the behavior of selected target audiences (TAs) toward US and multinational capabilities.

For further information on PSYOP support to FID, refer to JP 3-13.2, Psychological Operations.

(3) **Military Training to HN Forces**

(a) **US military training support to FID should focus on assisting HNs in anticipating, precluding, and countering threats or potential threats.** Emphasis on the HN's IDAD program when organizing, planning, and executing military training support helps the HN address the root causes of instability in a preventive manner rather than reacting to threats.

(b) **Security Force Assistance (SFA).** SFA is DOD's contribution to a unified action effort to support and augment the development of the capacity and capability of foreign security forces (FSF) and their supporting institutions to facilitate the achievement of specific objectives shared by the USG. The US military engages in activities to enhance the capabilities and capacities of a partner nation (or regional security organization) by providing training, equipment, advice, and assistance to those FSF organized in national ministry of defense (or equivalent regional military or paramilitary forces), while other USG agencies focus on those forces assigned to other ministries (or their equivalents) such as interior, justice, or intelligence services.

For additional discussion of SFA, see Chapter VI, "Foreign Internal Defense Operations," *paragraph 12,* "Military Training to Host Nation Forces."

(4) **Logistic Support.** US military capabilities may be used to **provide deployment and distribution, maintenance, supply, and construction support to the HN military or civilians** in operations that do not expose US personnel to hostile fire. The FAA does not generally authorize the transfer of equipment or supplies. Logistic support must be provided with consideration of the long-term effect on the capability of the local forces to become self-sufficient.

(5) **Intelligence Cooperation.** Intelligence cooperation is enabled by an information sharing environment that fully integrates joint, multinational, and interagency partners in a collaborative enterprise. US intelligence cooperation ranges from strategic analysis to current intelligence summaries and situation reporting for tactical operations. **Intelligence collection and dissemination capabilities are often weak links in an HN military capability.** US military communications hardware and operators may also be supplied in cases where HN infrastructure cannot support intelligence operations. The release of classified information to the HN is governed by national disclosure policy. Detailed written guidance may be supplemented with limited delegation of authority where appropriate.

d. **US Combat Operations. The introduction of US combat forces into FID operations requires a Presidential decision and serves only as a temporary solution until HN forces are able to stabilize the situation and provide security for the populace.** If combat is authorized, normally this will include major operations. In all cases, US combat operations support the HN IDAD program and remain strategically defensive in nature. While joint doctrine and Service tactics, techniques, and procedures provide guidance for specific operations and activities, there are certain principles that should guide employment of US forces conducting FID. These principles, and the specific C2 and employment considerations for joint and multinational tactical operations in FID, serve as the focus for discussions of tactical operations in this publication.

(1) The primary role for US military forces in tactical operations is to support, advise, and assist HN forces through logistics, intelligence or other combat support, and service support means. This allows the HN force to concentrate on taking the offensive against hostile elements.

(2) If the level of lawlessness, subversion, or insurgency reaches a level that HN forces cannot control, US forces may be required to engage the hostile elements with offensive operations in order to return the situation to a level controllable by HN forces. In this case, the objective of US operations is to protect or stabilize the HN political, economic, and social institutions until the host military can assume these responsibilities.

(3) In all cases, the strategic initiative and responsibility lie with the HN. To preserve its legitimacy and ensure a lasting solution to the problem, the host government must bear this responsibility. A decision for US forces to take the strategic initiative could amount to a transition to war.

(4) Given the multinational and interagency impact of conducting combat operations supporting FID, JFCs can expect complex C2 relationships. More information on C2 relationships and issues is provided in Chapter III, "Organization and Responsibilities for Foreign Internal Defense."

(5) The nature of US tactical participation in HN internal conflicts requires judicious and prudent rules of engagement (ROE) and guidelines for the application of force. Inappropriate destruction and violence attributed to US forces may easily reduce

the legitimacy and sovereignty of the supported government. In addition, these incidents may be used by adversaries to fuel anti-American sentiments and assist the cause of the opposition.

CHAPTER II
INTERNAL DEFENSE AND DEVELOPMENT

"Arguably the most important military component in the War on Terror[ism] is not the fighting we do ourselves, but how well we enable and empower our partners to defend and govern themselves. The standing up and mentoring of indigenous army and police—once the province of Special Forces—is now a key mission for the military as a whole."

Secretary of Defense Robert Gates, 26 November 2007

1. General

An IDAD program focuses on building viable political, economic, military, and social institutions that respond to the needs of the HN's society. Its fundamental goal is to prevent an insurgency or other forms of lawlessness or subversion by forestalling and defeating the threat and by working to correct conditions that prompt violence. The HN government mobilizes its population to participate in IDAD efforts. Thus, the IDAD program is ideally preemptive/Phase 0; however, if an insurgency, illicit drug, terrorist, or other threat develops, the IDAD program evolves to combat that threat. Commanders and their staffs must understand the HN's IDAD program and its objectives if they are to plan effectively to support it. The objectives of FID will be to assist the HN in formulating an appropriate IDAD program, which often includes fusing several separate strategic plans and programs into one broader strategy.

2. Construct

a. An IDAD program should integrate security force and civilian actions into a coherent, comprehensive effort. Security force actions provide a level of internal security that permits and supports growth through balanced development. This development requires change to meet the needs of vulnerable groups of people. This change may in turn promote unrest in the society. The strategy, therefore, includes measures to maintain conditions under which orderly development can take place.

b. Often a government must overcome the inertia and shortcomings of its own political system before it can cope with the internal threats it is facing. This may involve the adoption of reforms during a time of crisis when pressures limit flexibility and make implementation difficult.

c. The successful IDAD strategist must realize that the true nature of the threat to the government lies in the adversary's political strength rather than military power. Although the government must contain the armed elements, concentration on the military aspect of the threat does not address the real danger. Gaining support of the population is vital to any IDAD strategy. Any strategy that does not pay continuing, serious attention to the political claims and demands of the opposition is severely handicapped. Military and paramilitary programs are necessary for success, but are not sufficient alone.

3. Functions

a. An IDAD program blends four interdependent functions to prevent or counter internal threats (see Figure II-1). These functions are balanced development, security, neutralization, and mobilization.

b. **Balanced development** attempts to achieve national goals through political, social, and economic programs. It allows all individuals and groups in the society to share in the rewards of development, thus alleviating frustration. Balanced development

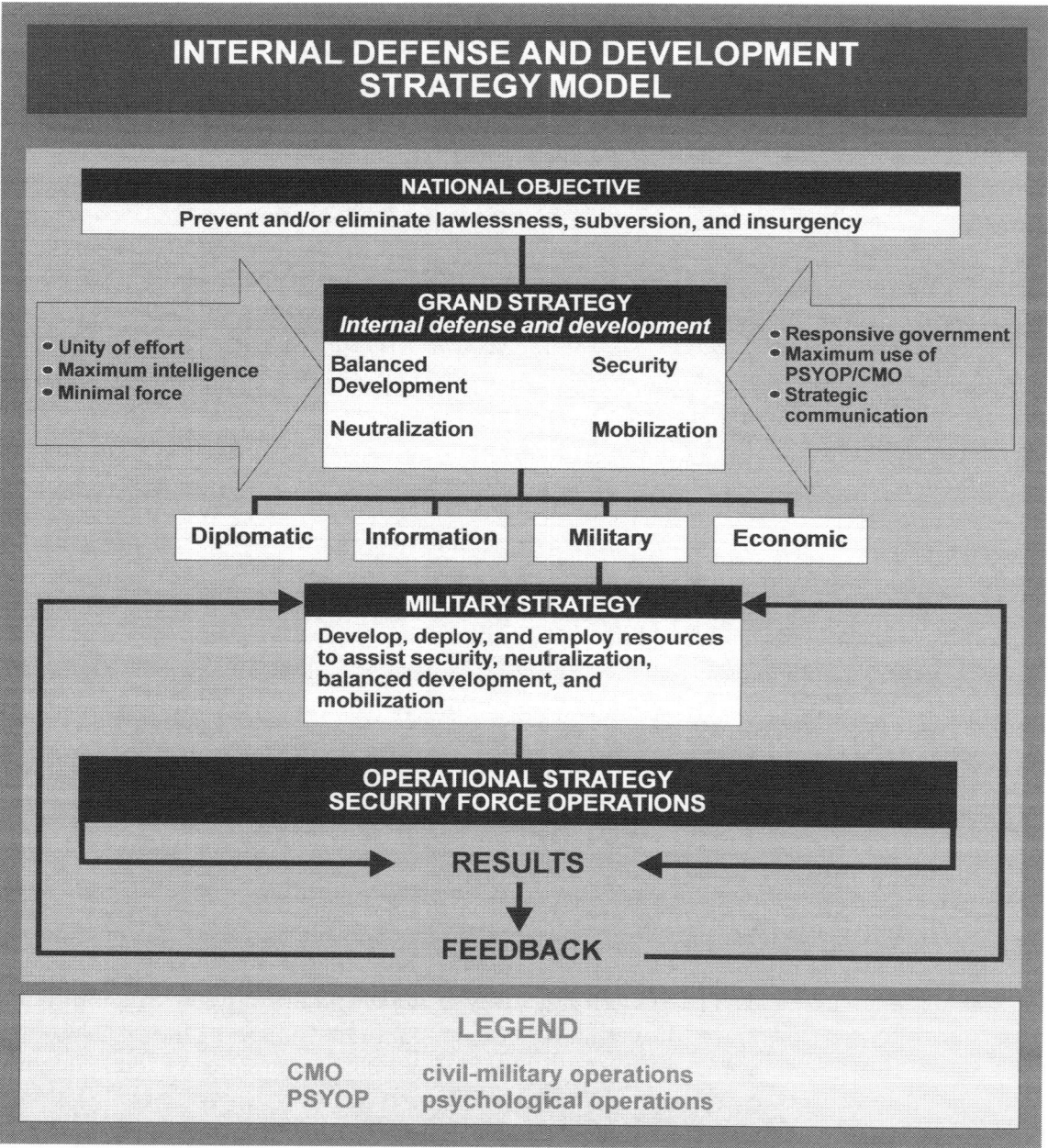

Figure II-1. Internal Defense and Development Strategy Model

satisfies legitimate grievances that the opposition attempts to exploit. The government must recognize conditions that contribute to the internal threat and instability and take preventive measures. Correcting conditions that make a society vulnerable is the long-term solution to the problem.

c. **Security** includes all activities implemented in order to protect the populace from violence and to provide a safe environment for national development. Security of the populace and government resources is essential to countering the threat. Protection and control of the populace permit development and deny the adversary access to popular support. The security effort should establish an environment in which the HN can provide for its own security with limited US support.

d. **Neutralization** is a political concept that:

(1) makes an insurgent or criminal element irrelevant to the political process;

(2) is the physical and psychological separation of the threatening elements from the population;

(3) includes all lawful activities (except those that degrade the government's legitimacy) to disrupt, preempt, disorganize, and defeat the insurgent organization;

(4) can involve public exposure and the discrediting of insurgent and criminal leaders during a period of low-level unrest with little political violence;

(5) can involve arrest and prosecution when laws have been broken; or

(6) can involve combat action when the adversary's violent activities escalate.

Note. All neutralization efforts must be legal. They must scrupulously observe HN laws and policy provisions regarding rights and responsibilities. The need for security forces to act lawfully is essential not only for humanitarian reasons but also because this reinforces government legitimacy while denying the adversary an exploitable issue. Special emergency powers may exist by legislation or decree. Government agents must not abuse these powers because they might well lose the popular support they need. Denying the adversary an opportunity to seize on and exploit legitimate issues against the government discredits their leaders and neutralizes their propaganda.

e. **Mobilization** provides organized manpower and materiel resources and includes all activities to motivate and organize popular support of the government. This support is essential for a successful IDAD program. If successful, mobilization maximizes manpower and other resources available to the government while it minimizes those available to the insurgent. Mobilization allows the government to strengthen existing institutions, to develop new ones to respond to demands, and promotes the government's legitimacy.

f. The HN continuously analyzes the results of its IDAD program, establishes measures of effectiveness (MOEs), and should have a methodology to provide feedback for future planning, refinement of strategy, and continued formulation of strategic national policy.

4. Principles

a. Although each situation is unique, certain principles guide efforts in the four functional areas to prevent or defeat an internal threat. Planners must apply the IDAD strategy and these principles to each specific situation. The principles are unity of effort, maximum use of intelligence, maximum use of CMO and PSYOP, minimum use of force, a responsive government, and use of strategic communication.

b. **Unity of Effort.** Unity of effort is the product of successful unified action and is essential to prevent crisis and defeat credible threats. Unity of effort requires cooperation among all forces and agencies toward a commonly recognized objective regardless of the command or coordination structures of the participants.

c. **Maximum Use of Intelligence.** Maximum use of intelligence requires that all operations be based on reliable, accurate, and timely intelligence. Successful implementation of operations necessitates an extensive operations security (OPSEC) and counterintelligence (CI) program to protect friendly FID operations and to counter and penetrate opposing force intelligence collection operations. Intelligence and CI operations must be designed so as to assess accurately the opposing force's capabilities, to provide timely indications and warning to HN and US forces, and to penetrate and be prepared to compromise hostile operations on order. If the HN is not capable of performing these missions effectively upon the commitment of US forces, then US intelligence and CI elements must be deployed to accomplish these missions. In these cases, FID operations may need to focus on support to HN efforts to develop its internal intelligence and security forces in order to perform these missions effectively. US elements may assist the HN in developing intelligence capability in accordance with USG directives and as deemed appropriate by the supported GCC in coordination with the US ambassador to the HN.

d. **Maximum use of CMO and PSYOP.** The effective use of CMO and PSYOP that are fully coordinated with other operational activities can enhance the legitimacy of HN forces and ultimately the stability of the HN. Stability operations, enabled by CMO, are the "deeds" that US and HN forces use to demonstrate and reinforce the "words," delivered through dedicated PSYOP and a broad based IO campaign. Use of CMO mitigates grievances exploited (or potentially exploited) by insurgents or other internal threats to HN stability by actively demonstrating an HN force's commitment to the well being of the population, and reinforces to key elements of the population their importance as the center of gravity and government's legitimacy. CMO is an initial step to reinforce and enhance the HN's image as a responsive government, both internally and internationally. The application of CMO integrated with PSYOP helps generate active and tacit popular support of the HN government and buys the time for the HN civil authorities and government to eliminate or mitigate valid popular grievances. CMO is executed by all forces and can be better enabled and facilitated by the application of CA forces to train, advise, and assist other forces, as well as plan and execute specific and targeted CAO to achieve operational and strategic objectives. Examples of coordinated CMO and PSYOP objectives in FID/IDAD include:

(1) Mitigate the grievances exploited by insurgents/threats to stability that generate popular support for resistance elements.

(2) Reduce the impact of military operations on the civil populace.

(3) Reduce civil interference with military operations.

(4) Facilitate civil order.

(5) Increase the effectiveness of, respect for, and cooperation with HN law enforcement and security forces.

(6) Set conditions and prepare the populace for elections.

e. **Minimum Use of Force.** A threatened government must carefully examine all COAs in response to internal violence. The government should stress the minimum use of force to maintain order and incorporate economy of force. At times, the best way to minimize violence is to use overwhelming force; however, normally the use of force should be appropriate (and proportional) to the incident at hand. In all cases where force is to be applied, clear rules on its use should be established, broadcast to the civilian populace, and understood by all members of the police, military, or paramilitary force being employed by the government.

f. **A Responsive Government.** Positive measures are necessary to ensure a responsive government whose ability to mobilize manpower and resources as well as to motivate the people reflects its administrative and management capabilities. In many cases, the leadership must provide additional training, supervision, controls, and follow-up.

g. **Use of Strategic Communication.** Strategic communication is an important element of strategic direction during all FID operations. DSPD, PA, and IO messages should be coordinated early during the planning process and continually throughout the operation. In FID operations, the use of strategic communication and PSYOP are closely related. Effectiveness and efficiency require the continual sharing of information. Although the messages may be different, they are derived from harmonized themes and must not contradict one another or the credibility of some of the agencies involved in the FID effort, particularly the HN's, could be compromised or lost.

5. Organizational Guidance

a. The following discussion provides a model for an organization to coordinate, plan, and conduct IDAD activities. Actual organizations may vary from country to country in order to adapt to existing conditions. Organizations should follow the established political organization of the nation concerned. The organization should provide centralized planning and direction and facilitate decentralized execution of the plan. The organization should be structured and chartered so that it can coordinate and direct the IDAD efforts of existing government agencies; however, it should minimize interference with those agencies' normal functions. Examples of national and subnational organizations show how to achieve a coordinated and unified effort at each level.

b. **National-Level Organization.** The national-level organization plans and coordinates programs. Its major offices normally correspond to branches and agencies of the national government concerned with insurgency, illicit drug trafficking, and terrorist or other internal threats. Figure II-2 depicts a planning and coordination organization at the national level.

(1) The planning office is responsible for long-range planning to prevent or defeat the threat. Its plans provide the chief executive with a basis for delineating authority, establishing responsibility, designating objectives, and allocating resources.

(2) The intelligence office develops concepts, directs programs, and plans and provides general guidance on intelligence related to national security. The intelligence office also coordinates intelligence production activities and correlates, evaluates, interprets, and disseminates intelligence. This office is staffed by representatives from intelligence agencies, police, and military intelligence.

(3) The populace and resource control office develops economy-related policies and plans, and provides general operational guidance for all forces in the security field. Representatives of government branches concerned with commerce, as well as law enforcement and justice, staff this office.

(4) The military affairs office develops and coordinates general plans for the mobilization and allocation of the regular armed forces and paramilitary forces. Representatives from all major components of the regular and paramilitary forces staff this office.

(5) Five separate offices covering PSYOP, information, economic affairs, cultural affairs, and political affairs represent their parent national-level branches or agencies, and develop operational concepts and policies for inclusion in the national plan.

(6) The duties of the administration offices are self evident and as directed.

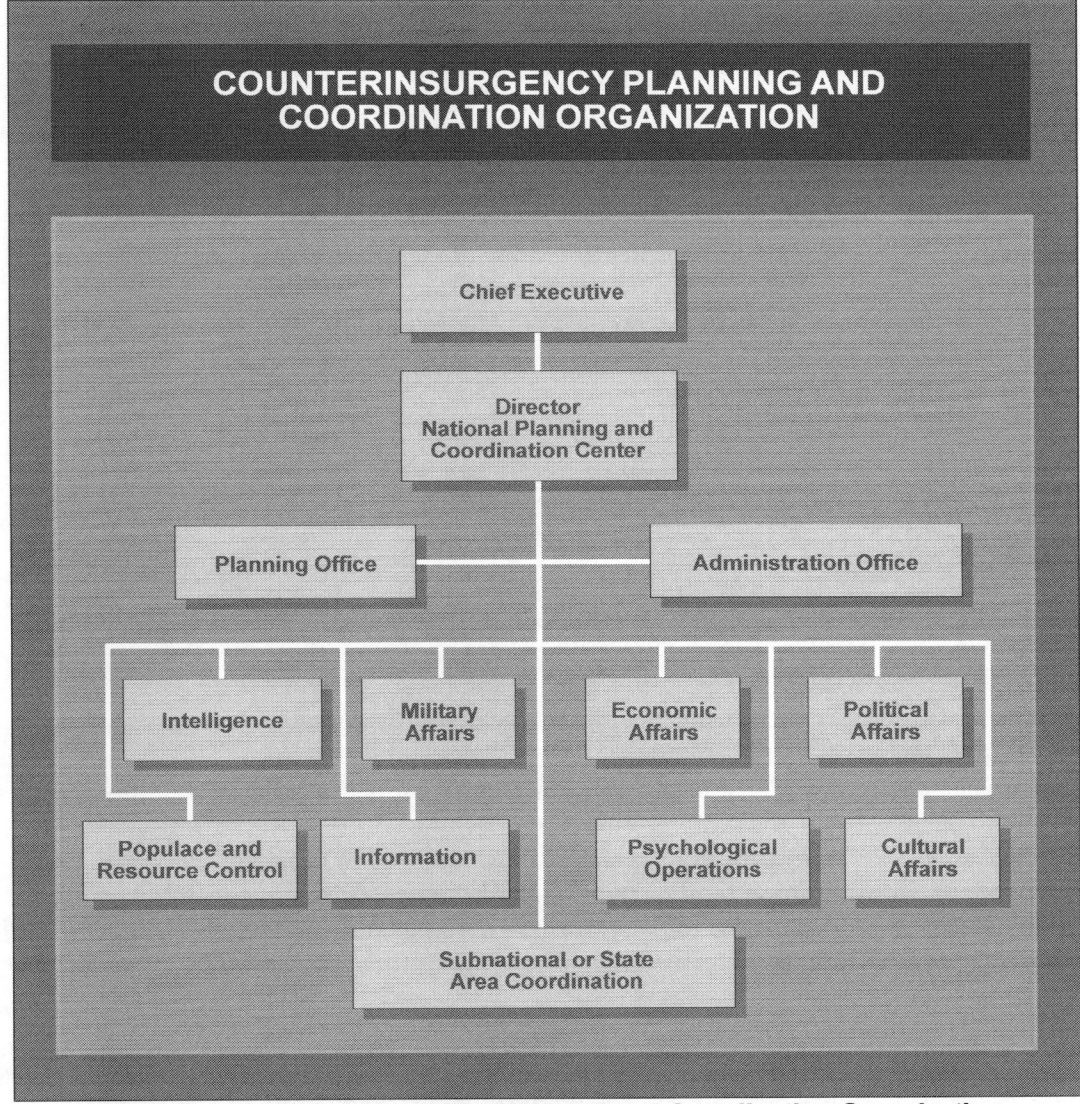

Figure II-2. Counterinsurgency Planning and Coordination Organization

c. **Subnational-Level Organization.** Area coordination centers (ACCs) may function as multinational civil-military headquarters at subnational, state, and local levels. ACCs plan, coordinate, and are staffed with commanders who exercise operational control (OPCON) over military forces made available. ACCs control civilian government organizations within their respective areas of jurisdiction. The ACC does not replace unit tactical operations centers or the normal government administrative organization in the operational area.

(1) ACCs perform a twofold mission: they provide integrated planning, coordination, and direction for all internal defense efforts and they facilitate an immediate, coordinated response to operational requirements. ACCs are headed by senior government officials who supervise and coordinate the activities of the staffs responsible for formulating internal defense plans and operations in their areas of interest.

The staffs contain selected representatives of major forces and agencies assigned to, or operating in, the center's area of interest. Each ACC includes members from the:

(a) Area military command.

(b) Area police agency.

(c) Local and national intelligence organization.

(d) Public information and PSYOP agencies.

(e) Paramilitary forces.

(f) Other local and national government offices involved in the economic, social, and political aspects of IDAD.

(2) There are two types of subnational ACCs that a government may form—regional and urban. The choice depends upon the environment in which the ACC operates.

(a) **Regional ACCs** normally locate with the nation's first subnational political subdivision with a fully developed governmental apparatus (state, province, or other). These government subdivisions are usually well established, having exercised government functions in their areas before the insurgency's onset. They often are the lowest level of administration able to coordinate all COIN programs. A full range of developmental, informational, and military capabilities may exist at this level. Those that are not part of the normal government organization should be added when the ACC activates. This augmentation enables the ACC to coordinate its activities better by using the existing structure.

(b) Urban areas usually require more complex ACCs than rural areas. **Urban ACCs** are appropriate for cities and heavily populated areas lacking a higher level coordination center. Urban ACCs are organized like the ACCs previously described and perform the same functions. However, the urban ACC includes representatives from local public service agencies, such as police, fire, medical, public works, public utilities, communications, and transportation.

(c) When a regional ACC resides in an urban area, economy of force and unity of effort may dictate that urban resources locate in that center where planners can coordinate and direct urban operations. The decision to establish an urban center or to use some other center for these purposes rests with the head of the government of the urban area who bases the decision on, among other things, available resources.

(d) If the urban area comprises several separate political subdivisions with no overall political control, the ACC establishes the control necessary for proper planning and coordination.

d. **Civilian Advisory Committees.** Committees composed of influential citizens help coordination centers at all levels monitor the success of their activities and gain popular support. These committees evaluate actions affecting civilians and communicate with the people. They provide feedback for future operational planning. Involving leading citizens in committees such as these increases their stake in, and commitment to, government programs and social mobilization objectives.

(1) The organization of a civilian committee varies according to local needs; changing situations require flexibility in structure. The chairman of the committee should be a prominent figure either appointed by the government or elected by the membership. General committee membership includes leaders in civilian organizations and other community groups who have influence with the target population. These leaders may include:

(a) Education officials (distinguished professors and teachers).

(b) Religious leaders.

(c) Health directors.

(d) Minority group representatives.

(e) Labor officials.

(f) Heads of local news media, distinguished writers, journalists, and editors.

(g) Business and commercial leaders.

(h) Former political leaders or retired government officials.

(i) Tribal or family leaders.

(j) Agricultural leaders, farmers, and stakeholders.

(2) The success of a civilian advisory committee hinges on including leading participants from all major political and cultural groupings, including minorities.

Intentionally Blank

CHAPTER III
ORGANIZATION AND RESPONSIBILITIES FOR
FOREIGN INTERNAL DEFENSE

> *"I believe it must be the policy of the United States to support free peoples who are resisting attempted subjugation by armed minorities or by outside pressures."*
>
> **Harry S. Truman,**
> **Message to Congress, 1947**

1. **General**

   a. **Integrated Effort**

      (1) When it is in the interests of national security, the United States may employ all the instruments of national power in order to assist friendly nations in conducting IDAD programs.

      (2) **For FID to be successful in meeting an HN's needs, the USG must integrate the efforts of multiple government agencies,** thus interorganizational coordination and cooperation becomes extremely important as it is the best way to integrate complementary efforts and effectively and efficiently use available resources. Effective integration is difficult and requires a consistent, focused effort that adjusts and evolves as the situation changes and different organizations and groups become involved.

      (3) Such integration and coordination are essentially vertical between levels of command and organization, and horizontal between USG agencies and HN military and civilian agencies. In addition, integration and coordination requirements may extend to allied nations and coalition partners participating with the US in multinational FID efforts. As Figure III-1 illustrates, **the lines of organization and coordination during FID operations are complex.** This factor, combined with the breadth of potential FID operations, makes complete integration and coordination of all national FID efforts a daunting challenge.

   b. **Organizing for FID Operations**

      (1) Management of the FID effort begins at the national level, with the selection of those nations the US will support through FID efforts. This decision is made by the President with advice from the Secretary of State, SecDef, and other officials. Funding for these programs is appropriated by Congress. The US will consider FID support when the following three conditions exist:

(a) The existing or threatened internal disorder threatens US national strategic goals.

(b) The threatened nation is capable of effectively using US assistance.

(c) The threatened nation requests US assistance.

(2) The level and type of assistance required is determined and a country-specific plan is developed. No two FID operations are exactly alike.

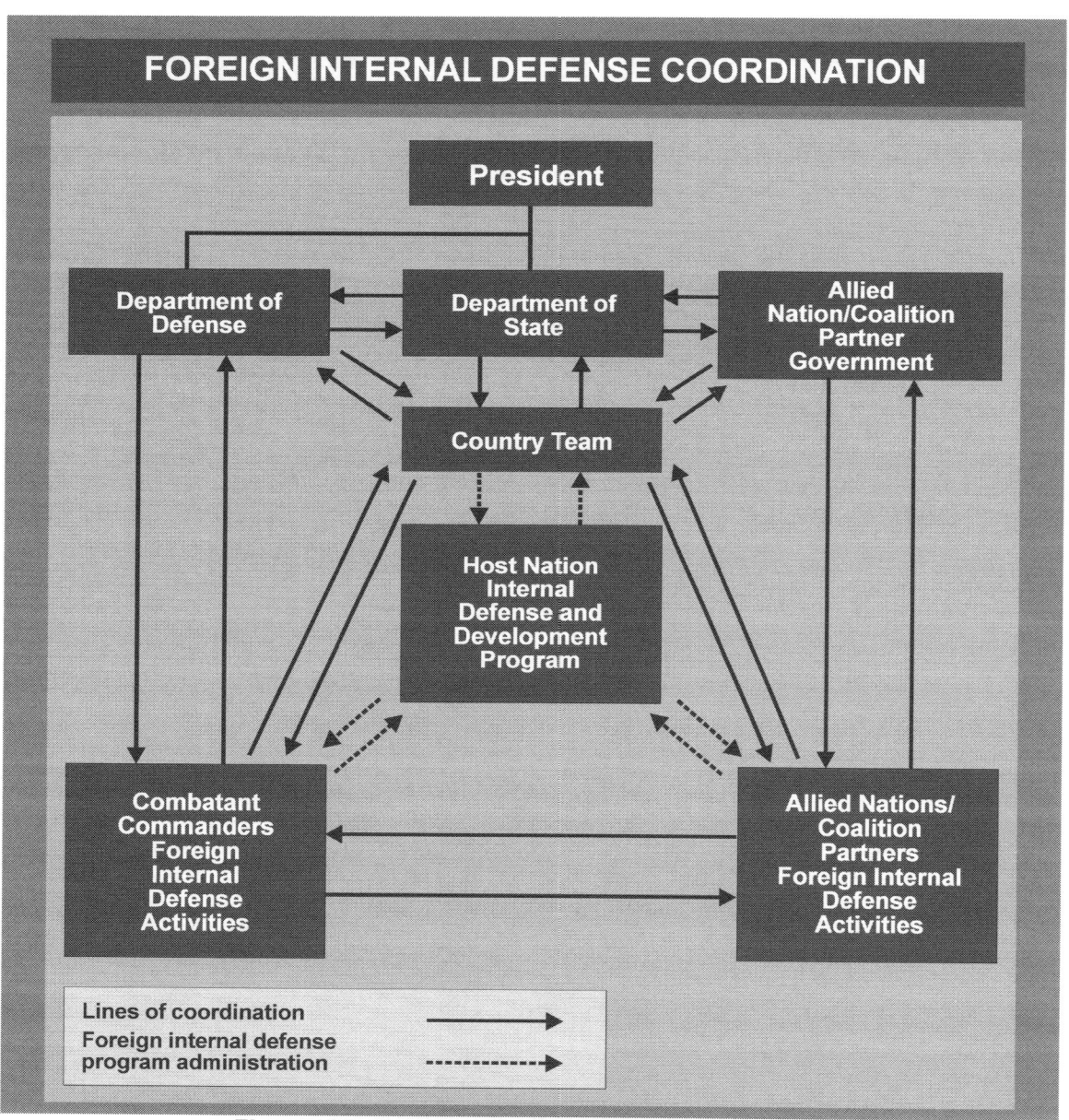

Figure III-1. Foreign Internal Defense Coordination

(3) The GEF describes SecDef's priorities for creating new partnerships and building the capacity of existing partnerships. The joint force conducts FID in compliance with this guidance.

(4) Ordinarily, when the decision limits FID support to minor levels of SA or CMO, there is no requirement to establish a special management program to facilitate interagency coordination. In these cases, standard interagency coordination should be adequate. The major FID efforts (i.e., those in support of nations important to US national interests) demand levels of management and coordination beyond what is normally found at the interagency, combatant command, and country team levels.

2. National-Level Organizations

The National Security Council (NSC) will generally provide the initial guidance and translation of national-level decisions pertaining to FID. Specific guidance will also be provided to government agencies and departments. Key government agencies that normally take part in FID are:

a. **Department of State. DOS is generally the lead government agency in executing US FID operations.** Major responsibilities of DOS related to the planning and execution of FID include:

(1) The Secretary of State has responsibility for advising the President in forming foreign policy and has other major specific responsibilities key to the overall planning and execution of the national FID effort.

(2) DOS assists the NSC in building national FID policies and priorities, and is the lead government agency to carry out these policies in the interagency arena. DOS involvement extends from policy formulation at the highest level to mission execution at the HN and country team levels. The Policy Planning Staff, Bureau of Political-Military Affairs (PM), and the Office of the Coordinator for Reconstruction and Stabilization are the elements of DOS most involved with interagency planning for FID operations. The Bureau of Finance and Operations provides financial and operational support services to all of the department's program areas.

(3) The Under Secretary of State for Arms Control and International Security is the principal advisor and focal point for SA matters within DOS. Control and coordination of SA extends from this office to the Assistant Secretary of State for Political-Military Affairs. These lines of supervision and administration interface with DOD at the individual country teams and security cooperation organizations (SCOs) in the HNs (see Figure III-1). DOS directs the overall US SA program; DOD executes the program.

(4) At the national level, the DOS PM is the principal channel of liaison between DOS and DOD. This bureau has primary responsibility for assisting the Secretary of State in executing the responsibilities of managing the military portion of

SA. PM is also the focal point within DOS for providing policy direction in international security, military operations, defense strategy and plans, and defense trade.

(5) INL (DOS) is especially important in FID operations as a coordinating link in US support of HN CD programs.

(6) The Bureau of International Information Programs (IIP) supports US foreign policy objectives by informing the public in other nations about US programs and policies, and administering overseas cultural and exchange programs. These activities enhance US military operations in support of HN IDAD programs through public diplomacy to the supported government and its populace. DOD and DOS IIP efforts must be mutually supportive. Close coordination among embassy public affairs officers (PAOs) and cultural attachés, military PA offices, and PSYOP elements is essential.

(7) **United States Agency for International Development (USAID).** Although USAID and DOS are separate organizations, both report to the Secretary of State. USAID activities have a significant impact on military activities in support of FID. **USAID carries out nonmilitary assistance programs designed to assist certain less developed nations to increase their productive capacities and improve their quality of life.** It also promotes economic and political stability in friendly nations. The mission of USAID and the parallel DOD developmental activities supporting FID underscore the importance of employing an integrated interagency effort.

b. **Director of National Intelligence (DNI) and the Director of the Central Intelligence Agency (D/CIA).** The DNI and D/CIA support the FID mission in both a national-level advisory capacity and at the regional and country levels through direct support of FID activities. The DNI advises the NSC in matters concerning the coordination and implementation of intelligence activities in support of national-level FID efforts. On the regional level, the Central Intelligence Agency (CIA) provides intelligence in support of FID threat analysis and needs assessments and supports the chief of mission (COM) with intelligence at the country team level. This intelligence support is extremely important in determining the level and degree of required resources and in determining the effectiveness of these committed resources. **Military intelligence activities are linked with CIA activities, either directly or through the country team**, to ensure the exchange of information necessary to support FID.

c. **Department of Defense.** The DOD national-level organizations involved in FID management include the Office of the Secretary of Defense (OSD) and the Joint Staff.

(1) **Office of the Secretary of Defense.** In most FID matters, **OSD acts as a policy-making organization.** Numerous activities at the OSD level affect FID efforts. The five activities listed below are directly involved in the areas of SA and in the general areas of low-intensity conflict- and FID-related issues.

(a) The Under Secretary of Defense for Policy (USD[P]) exercises overall direction, authority, and control concerning SA for OSD through the various assistant secretaries of defense.

(b) The Assistant Secretary of Defense (Special Operations and Low-Intensity Conflict and Interdependent Capabilities) (ASD[SO/LIC&IC]) oversees DOD special operations (SO) and has far-reaching policy responsibilities that can impact on virtually all areas of FID policy and programs.

(c) The Assistant Secretary of Defense (Global Security Affairs) establishes SA policy and supervises SA programs through the Defense Security Cooperation Agency (DSCA).

(d) DSCA is the principal DOD organization through which SecDef carries out responsibilities for SA. DSCA administers and supervises SA planning and formulates and executes SA efforts in coordination with other government programs. DSCA also conducts international logistics and sales negotiations with representatives of foreign nations and serves as the DOD focal point for liaison with US industry regarding SA. Finally, DSCA develops and promulgates SA procedures, maintains the database for the programs, and makes determinations with respect to the allocation of FMS administrative funds.

(e) The Assistant Secretary of Defense for Public Affairs supervises and establishes policy for PA programs with DOD. PA is an integral part of military support to FID.

(2) **Chairman of the Joint Chiefs of Staff (CJCS).** The CJCS **plays an important role in providing strategic guidance to the combatant commanders (CCDRs) for the conduct of military operations to support FID.** This guidance is provided primarily through the NMS and the Joint Strategic Capabilities Plan (JSCP), the key components of the Joint Strategic Planning System (JSPS). This guidance is provided after, and often modified as a result of, the interagency coordination and policy development process described earlier in this chapter. Because of their familiarity with the needs of the friendly nations in their regions, the GCCs are given great latitude in managing and coordinating their military activities in support of FID.

d. **United States Coast Guard, within the Department of Homeland Security.** The USCG is specifically authorized to assist other federal agencies in the performance of any activity for which the USCG is especially qualified, including SA activities for DOS and DOD. The *Memorandum of Agreement Between the Department of Defense and the Department of Homeland Security on the Use of US Coast Guard Capabilities and Resources in Support of the National Military Strategy* identifies certain national defense capabilities of the USCG, including theater SC, and improves the process by which the USCG serves as a force provider for DOD missions.

3. Combatant Commands

a. **GCCs** are responsible for planning and executing military operations in support of FID within their area of responsibility (AOR). **Other CCDRs** play a supporting role by providing resources to conduct operations as directed by the President or SecDef.

b. The GCC has the responsibility of coordinating and monitoring all the military activities in the AOR in support of FID. **GCCs** develop TCPs that include SC programs and activities in accordance with the GEF. Organizing for military operations in FID will vary, but there are fundamental principles that apply when planning or executing FID operations. For example:

(1) Military activities in support of FID are an integral part of the long-range strategic plans and objectives for the GCC's AOR. These plans must reflect national security priorities and guidance.

(2) GCCs may request to expand the military presence in the country team. In most instances, the application of US military resources in support of an HN's IDAD programs will function through the framework of SCOs. However, should it become necessary to expand US assistance by introducing selected US military forces, a JTF or JSOTF normally will be established to coordinate this effort.

c. **Staff Organization.** The general purpose and functions of the CCDR's joint staff are provided in JP 1, *Doctrine for the Armed Forces of the United States,* and JP 3-33, *Joint Task Force Headquarters.* The purpose of this discussion is to outline general organizational requirements for FID oversight and management at the combatant command level.

(1) **Plans Directorate.** The plans directorate of a joint staff (J-5), as the staff planner, **incorporates military support to FID into theater strategy and plans.** The J-5 has three ways to accomplish this: the plans division prepares the GCC's vision and strategy, looking out 5 to 10 years and providing long-term and mid-term objectives for military support to FID; the political-military affairs division links the combatant command to the SCOs; and the SA section provides oversight of military SA efforts and coordinates integration of regional SA activities into theater-wide activities. The SA section may be organized within another directorate of the combatant command staff, depending on the desires of the GCC.

(2) **Operations Directorate.** The operations directorate of a joint staff (J-3) **monitors the execution of military operations in support of FID.** Additionally, the J-3 may use the CMO and the PSYOP sections to orchestrate specifically designed programs to maximize the positive effects of military activities in support of FID. The J-3 also employs an SO staff element that assists in the planning and employment considerations for SOF in support of FID. GCCs may elect to assign the above programs and planning for SOF in support of FID to the theater subunified special operations commander (SOC) or designated element.

(3) **Intelligence Directorate. The intelligence directorate of a joint staff (J-2) produces intelligence that often supplements estimates produced by the national intelligence agencies.** The J-2 leads the staff's effort to conduct joint intelligence preparation of the operational environment (JIPOE). Intelligence requirements (IRs) in support of FID include assessment of economic, political, and social conditions; as well as the accurate detection of internal instability. Through the counterintelligence support officer (CISO) the J-2 is also responsible for planning for CI support to the military portion of the FID operation. CI support is critical to the execution of FID operations and can provide commanders with valuable tools for force protection (FP) planning as well as maintain the integrity and OPSEC of FID operations. Additionally, in those instances where a force protection detachment (FPD) is stationed in the HN, the CISO or designated CI coordinating authority is responsible for coordinating FID support. Cooperative intelligence liaisons between the US and the HN are vital; however, disclosure of classified information to HN or other multinational FID forces must be authorized. CI support needs to be included and executed from the initiation of the FID operation. Due to the nature of CI, it cannot be relied upon to be fully effective or productive if included as an afterthought or brought into action after FID operations are underway.

(4) **Political Advisor (POLAD).** GCCs may be assigned a POLAD by DOS. **The POLAD serves as a link between DOS and the GCC's staff.** An effective use of the POLAD's skill in FID-intensive theaters may be for the GCC to establish **a FID interagency working group** (see Figure III-2) consisting of interagency representatives and military staff from country teams and the GCC's staff. This group acts as a focal point for the coordination and integration of military and nonmilitary support to FID.

(5) **Legal Advisor.** The legal advisor should evaluate all military operations in support of FID because of the legal restrictions and complex funding sources. See Appendix A, "Legal Considerations," for additional considerations when supporting FID.

(6) **Public Affairs.** Integral to successful military operations in support of FID is public awareness and support. A coordinated public information program to support a FID operation is essential. The PAO must be an early and active participant in military planning to support FID.

(7) When formed, a **joint interagency coordination group (JIACG)** can provide the GCC with an increased capability to collaborate FID operations with other USG civilian agencies and departments.

For more information on JIACGs, see JP 3-08, Interorganizational Coordination During Joint Operations.

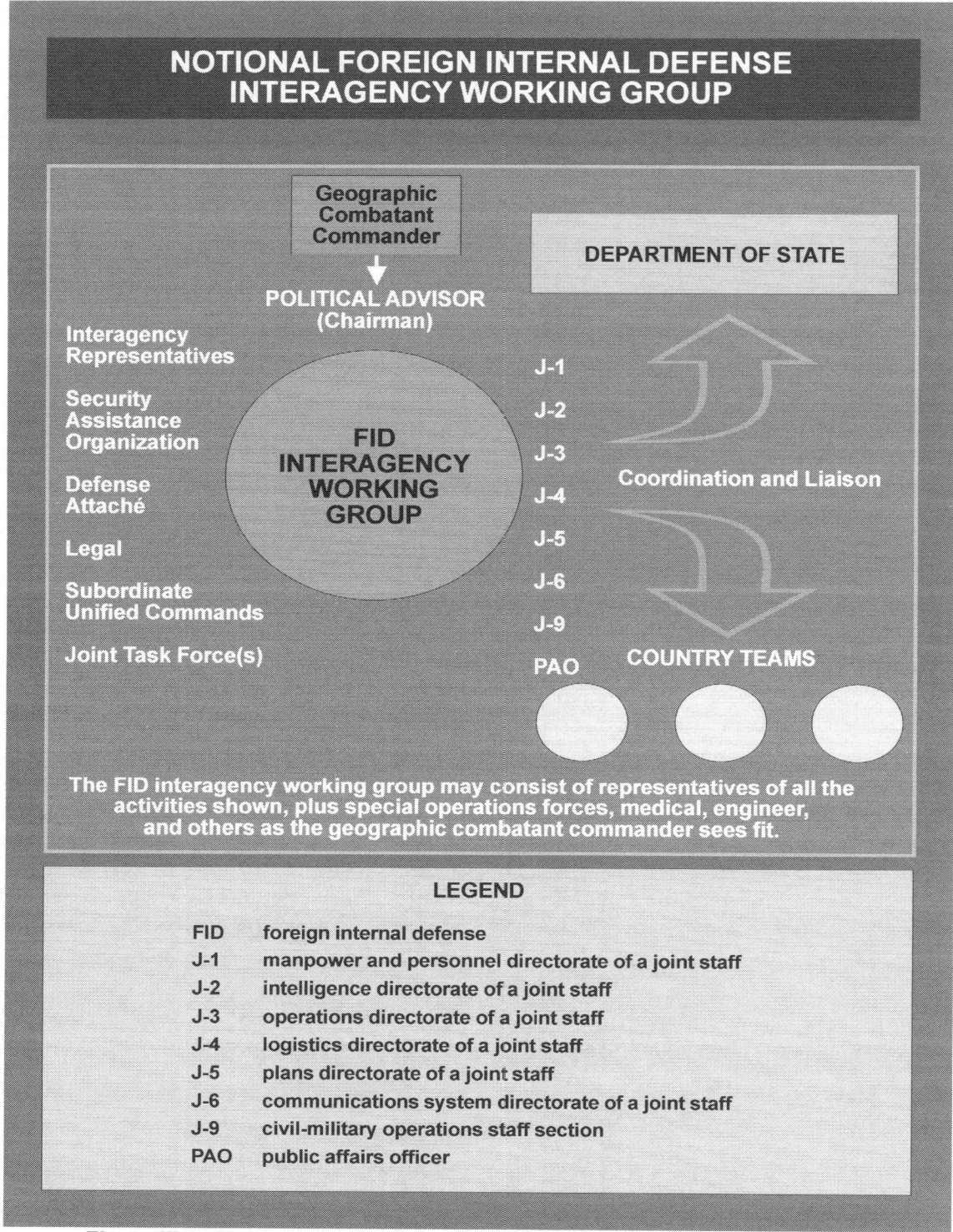

NOTIONAL FOREIGN INTERNAL DEFENSE INTERAGENCY WORKING GROUP

Geographic Combatant Commander

DEPARTMENT OF STATE

POLITICAL ADVISOR (Chairman)

Interagency Representatives

Security Assistance Organization

Defense Attaché

Legal

Subordinate Unified Commands

Joint Task Force(s)

FID INTERAGENCY WORKING GROUP

J-1
J-2
J-3
J-4
J-5
J-6
J-9
PAO

Coordination and Liaison

COUNTRY TEAMS

The FID interagency working group may consist of representatives of all the activities shown, plus special operations forces, medical, engineer, and others as the geographic combatant commander sees fit.

LEGEND

FID   foreign internal defense
J-1   manpower and personnel directorate of a joint staff
J-2   intelligence directorate of a joint staff
J-3   operations directorate of a joint staff
J-4   logistics directorate of a joint staff
J-5   plans directorate of a joint staff
J-6   communications system directorate of a joint staff
J-9   civil-military operations staff section
PAO   public affairs officer

Figure III-2. Notional Foreign Internal Defense Interagency Working Group

(8) **Other Staff Elements.** All staff elements contribute to the overall support of the FID operation. Some, such as the logistics directorate of a joint staff (J-4) and the communications system directorate of a joint staff, may be given primary responsibility

for specific military technical support missions. These staff elements will usually focus on the direct support (not involving combat operations) category of military support to FID.

4. Subordinate Unified Commands

a. When authorized by SecDef through the CJCS, commanders of unified commands may establish subordinate unified commands (also called subunified commands) to conduct operations on a continuing basis in accordance with the criteria set forth for unified commands.

b. A subordinate unified command (e.g., United States Forces Korea) may be established on a geographical area or functional basis. The responsibilities for FID support in these commands closely parallel those discussed for the combatant commands. Specific authority for planning and conducting FID depends on the level of authority delegated by the CCDR. However, basic principles and staff organization remain consistent.

c. Functional subordinate unified commands such as Special Operations Command South, which is the theater special operations command (TSOC) for United States Southern Command (USSOUTHCOM), control a specific functional capability. These functional commands contribute to FID planning and execution through management of FID areas related to their functional areas of expertise.

d. **TSOCs are of particular importance because of the significant role of SOF in FID operations.** The theater SOC normally has OPCON of SOF in the theater and has primary responsibility to plan and execute SOF operations in support of FID. SOF assigned to a theater are under the combatant command (command authority) of the GCC. The GCC normally exercises this authority through the commander of the TSOC. When a GCC establishes and employs multiple JTFs and independent task forces, the TSOC commander may establish and employ multiple JSOTFs to manage SOF assets and accommodate JTF/task force SO requirements. Accordingly, the GCC, as the common superior, normally will establish supporting or tactical control command relationships between JSOTF commanders and JTF/task force commanders. Coordination between the joint force SO component commander (who is also the TSOC commander) and the other component commanders within the combatant command is essential for effective management of military operations in support of FID, including joint and multinational exercises, mobile training teams (MTTs), integration of SOF with CF, and other operations.

For further information on JSOTFs, refer to JP 3-05.1, Joint Special Operations Task Force Operations.

5. Joint Task Forces

GCCs may form JTFs to execute complex missions. For example, United States Northern Command's Joint Task Force North provides US military assistance to US civil law enforcement agencies to combat transnational threats to the homeland. Another example, JTF-BRAVO, which is subordinate to USSOUTHCOM, was formed by the CCDR for the primary mission of coordinating and supporting US military training exercises in Honduras during a time when a US forward presence in Central America was deemed necessary. The large number of training exercises and related HCA projects conducted were a primary factor in the decision to form the JTF. Other JTFs may be organized to accomplish specific functional missions such as road construction and support for transportation and communications efforts. Much of the training, CAO, and PSYOP conducted by a JTF may warrant the creation of a subordinate JSOTF, a joint CMO task force, or a joint PSYOP task force. In some instances a JSOTF may be the initial or follow-on task force structure when conducting FID operations. As seen in recent experience during Operation ENDURING FREEDOM-Philippines, the JSOTF structure may include a subordinate CF element.

For further information on JTFs, refer to JP 3-33, Joint Task Force Headquarters.

6. The United States Diplomatic Mission and Country Team

a. The US diplomatic mission to an HN includes representatives of all US departments and agencies physically present in the country. **The President gives the chief of the diplomatic mission, normally an ambassador, full responsibility for the direction, coordination, and supervision of all USG executive branch employees in-country.** The COM has authority over all USG executive branch employees within the mission and host country except for employees under the command of a US military commander (Title 22, USC, Section 3927). However, this authority does not extend to personnel in other missions or those assigned to an international agency. Close coordination with each COM and country team in the GCC's AOR is essential in order to conduct effective, country-specific FID operations that support the HN's IDAD program and US regional goals and objectives.

b. **Organization.** The **country team structure** (see Figure III-3) denotes the process of in-country, interdepartmental coordination among key members of the US diplomatic mission. The composition of a country team varies, depending on the desires of the COM, the in-country situation, and the number and levels of US departments and agencies present. **The principal military member of the country team is the senior defense official/defense attaché (SDO/DATT).** In addition to being the diplomatically accredited DATT the SDO is the chief of the SCO. In some instances the COM may use the term security assistance organization/officer for the organization or his or her staff members. Although the US area military commander (the GCC or a subordinate) is not a member of the diplomatic mission, the commander may participate or be represented in meetings and coordination conducted by the country team. The following discussion

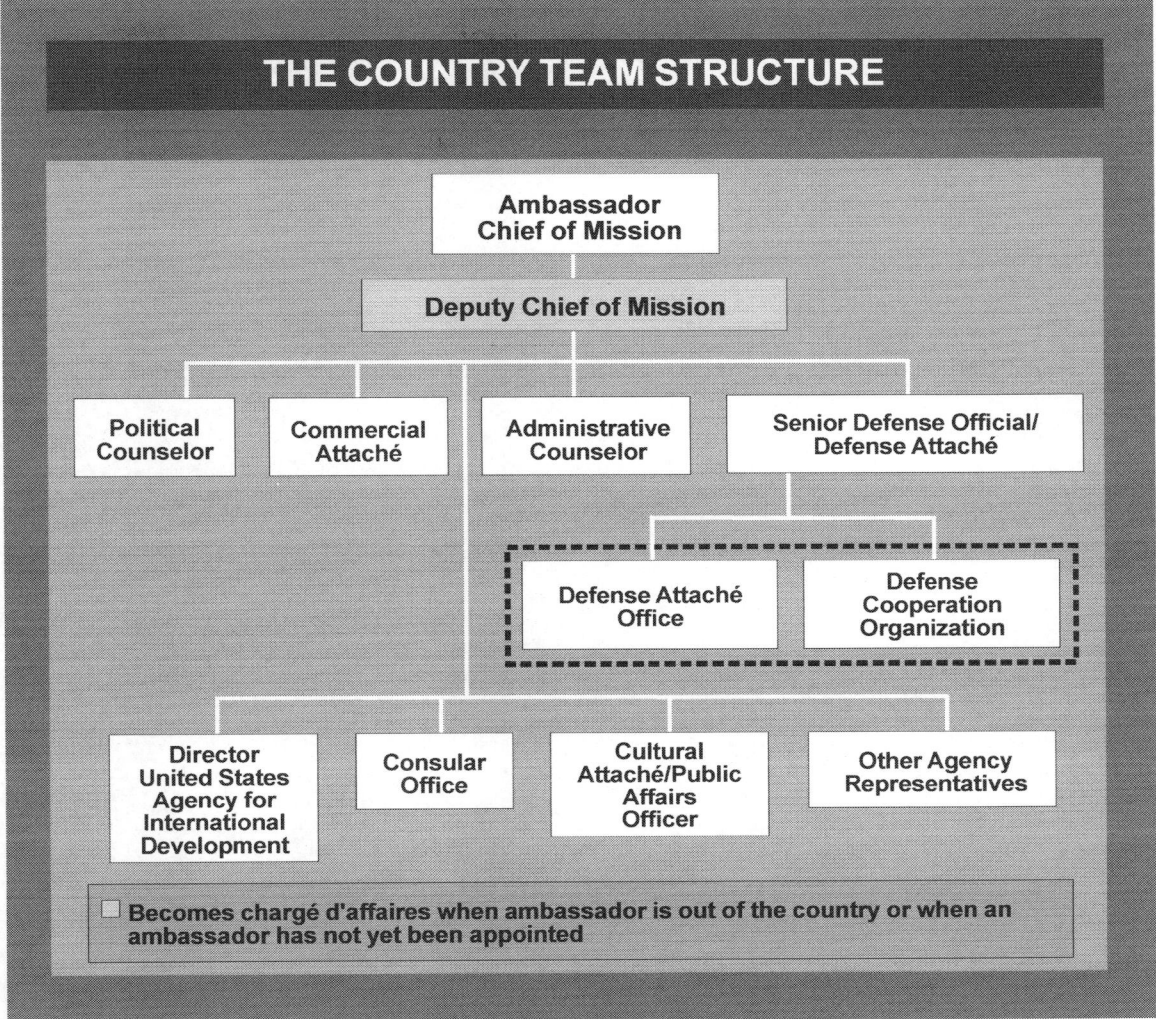

Figure III-3. The Country Team Structure

provides an outline of typical country team representatives and explains the military elements important to the FID mission.

(1) The COM coordinates much of the FID effort in the assigned country and accomplishes this task either through the assigned SCO or through the country team.

(2) DOS is generally represented on the country team by the following:

(a) The deputy COM serves as executive officer and chief of staff for the ambassador and directs the diplomatic mission in the ambassador's absence (then called the chargé d'affaires).

(b) The political counselor directs the political section and is often third in command of the mission. The political section may also contain a political and/or military officer to assist in the coordination of military activities supporting FID operations.

(c) The commercial attaché is trained by the Department of Commerce and promotes US commercial interests.

(d) The cultural attaché is a DOS public diplomacy officer responsible for implementing the US information program throughout the HN. This individual will often double as the PAO.

(3) USAID is represented by the in-country director of USAID. The director directs the nonmilitary US developmental efforts in the HN.

(4) Other USG departments, agencies, and interests may be represented by the following:

(a) Treasury attaché.

(b) Agricultural attaché.

(c) Labor attaché.

(d) Air attaché.

(e) Science attaché.

(f) DEA representative.

(g) Director of the Peace Corps.

(h) Legal attaché (representing DOJ).

(i) USCG representative.

(5) The DOD organization and representation within the diplomatic mission and country team can range from as little as an envoy to a full complement of Service attachés or a major SCO. In nations with active FID operations, there is likely to be a larger military presence with most of these resources centered in the SCO.

(a) The SDO/DATT is the principal DOD official in a US embassy, as designated by SecDef. The SDO/DATT is the COM's principal military advisor on defense and national security issues, the senior diplomatically accredited DOD military officer assigned to a US diplomatic mission, and the single point of contact for all DOD matters involving the embassy or DOD elements assigned to or working from the embassy. All DOD elements assigned or attached to or operating from US embassies are aligned under the coordinating authority of the SDO/DATT. Where separate SC and DATT offices exist, they remain separate with distinct duties and statutory authorities. The SDO/DATT represents SecDef and the appropriate GCC for coordination of administrative and security matters for all DOD personnel not under the command of a

US commander. The SDO/DATT in each embassy exercises coordinating authority over DOD elements under COM authority, subject to the authorities of the GCC and DOD component commanders. This coordinating authority does not preempt the authority exercised over these elements by the COM, the mission authority exercised by the parent DOD components, or the command authority exercised by the GCC under the Unified Command Plan. As required by the COM and with concurrence of the GCC, the SDO/DATT assumes tactical control of DOD elements assigned or attached to the embassy for country team support to significant embassy events.

(b) **The SDO/DATT is the officer in charge of the US defense attaché office (USDAO).** The SDO/DATT and other Service attachés serve as SecDef's, CJCS's, and CCDR's diplomatic representatives to their HN counterparts. USDAOs are operated by Defense Intelligence Agency. The attachés also serve the ambassador and coordinate with and represent their respective Military Department on Service matters. The attachés assist the FID effort by exchanging information with the CCDR's staff on HN military, social, economic, and political conditions. In the majority of countries, the functions of an SCO are carried out within the USDAO under the direction of the SDO/DATT.

(c) In nations where large concentrations of US military personnel are resident or transit, FPDs may be deployed to support DOD CI. The primary mission of an FPD is to provide AT/FP indications and warnings to DOD personnel. FP personnel are specially trained, area-oriented, and language-qualified. FPDs interface with HN local law enforcement and security services to enhance CI and security support to FP for US troops in theater. OSD-sponsored FPDs may also be deployed to locations where DOD military criminal investigative organizations are not currently present.

(d) **The SCO is the most important FID-related military activity under the supervision of the ambassador.** The specific title of the SCO may vary; however, these differences reflect nothing more than the political climate within the HN. As examples, an SCO may be referred to as a military assistance advisory group, military advisory group, office of military cooperation, or office of defense cooperation. SCOs may have up to six members of the Armed Forces before congressional approval is required. Usually, a US military officer serves as the chief. When programs involve more than one Service, the SCO organization will be joint. The organization (departmental and functional alignments) of a typical SCO are indicated in Figures III-4 and III-5.

1. The chief, SCO (sometimes referred to as the security assistance officer) operates under the direction of the US ambassador (COM) and reports administratively to the GCC and is funded by DSCA. The SCO assists HN security forces by planning and administering military aspects of the SA program. The SCO also helps the US country team communicate HN assistance needs to policy and budget officials within the USG and combatant command for recommended inclusion in the Foreign Operations Budget.

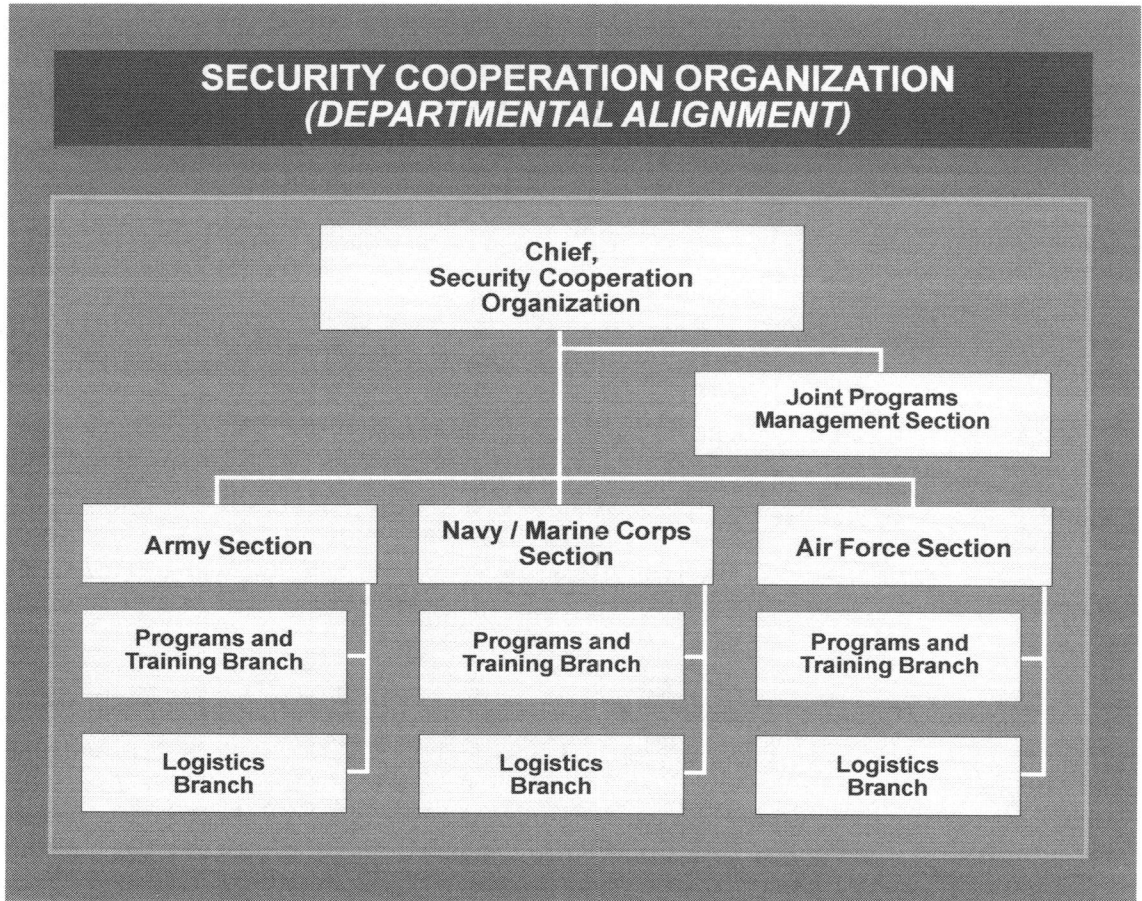

Figure III-4. Security Cooperation Organization (Departmental Alignment)

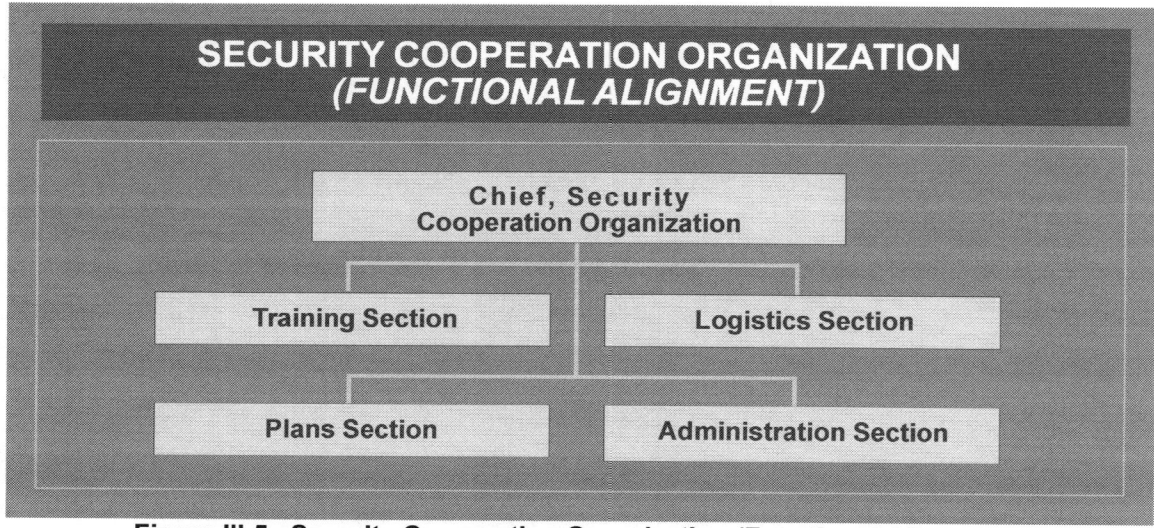

Figure III-5. Security Cooperation Organization (Functional Alignment)

2. The SCO is essentially a management organization that helps assess the HN needs and articulate them through the instruments described above. SCOs manage equipment and service cases; manage training; monitor programs; evaluate and plan HN military capabilities and requirements; provide administrative support; promote rationalization, standardization, and interoperability; and perform liaison exclusive of advisory and training assistance. In addition, the SCO provides oversight of training and assistance teams temporarily assigned to assist the HN. The SCO is limited by law from giving direct training assistance that is normally provided through special teams and organizations assigned to perform limited tasks for specific periods. These include technical assistance field teams (TAFTs), MTTs, technical assistance teams (TATs), language training detachments, weapon system logistics offices, quality assurance teams (QATs), as well as site survey and defense requirement survey teams. The SDO/DATT has the command responsibility to expeditiously determine the status and whereabouts of all assigned or attached DOD-affiliated personnel in conjunction with a natural or man-made disaster and accomplish reporting. The GCC's manpower and personnel directorate of a joint staff (J-1) will establish the actual report format and direct submission timelines and procedures.

(6) Extensive and effective lines of coordination in an organization such as the country team are critical to its effective functioning. Effective coordination from the national level down to the smallest independent agencies operating within the HN is essential. This arrangement and lines of coordination are illustrated in Figure III-6.

7. Multinational Foreign Internal Defense Force

a. **Multinational operations require clear C2 and coordination procedures for FID planning and execution in order to facilitate unity of effort.**

b. Each multinational operation in support of FID is different, and key considerations involved in planning and execution may vary with the international situation and perspectives, motives, and values of the organization's members.

c. **Multinational Resources.** Multinational partners can assist materially in training HN security forces. Some nations more willingly train HN forces, especially police forces, than provide troops for combat operations. Some multinational forces come with significant employment restrictions. Each international contribution is considered on its own merits, but such assistance is rarely declined. Good faith efforts to integrate multinational partners and achieve optimum effectiveness are required.

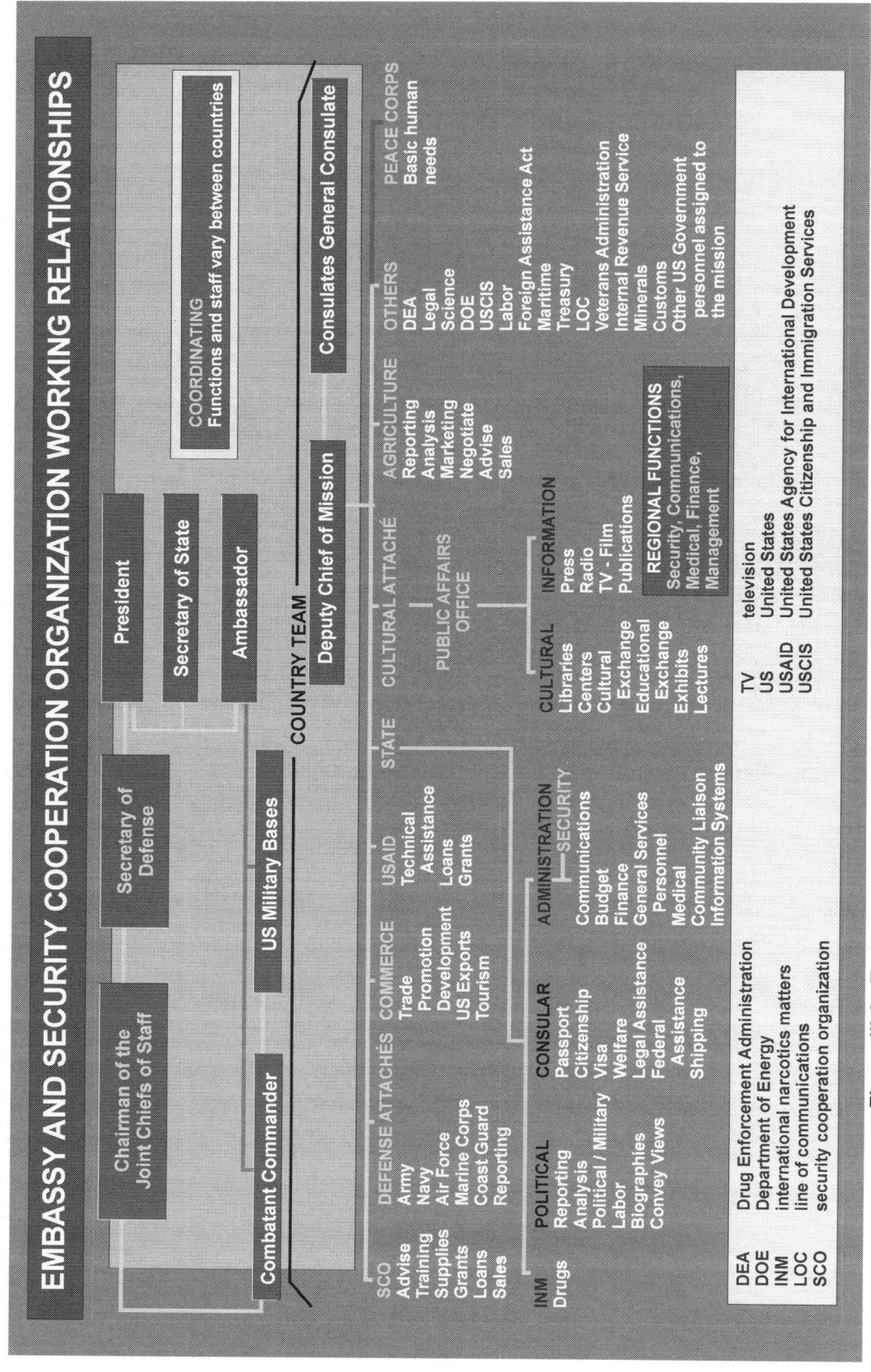

Figure III-6. Embassy and Security Cooperation Organization Working Relationships

CHAPTER IV
PLANNING FOR FOREIGN INTERNAL DEFENSE

> *"The security of the United States is tightly bound up with the security of the broader international system. As a result, our strategy seeks to build capacity of fragile or vulnerable partners to withstand internal threats and external aggression while improving the capacity of the international system itself to withstand the challenge posed by rogue states and would-be hegemons."*
>
> **National Defense Strategy, June 2008**

1. General

a. FID is designed to bolster the internal stability and security of the supported nation. Only a comprehensive planning process at both the national and theater level can provide the means to reach this goal.

b. **NSC Directives Promulgate US FID Policy.** JSPS documents reflect the military responsibilities for carrying out this broad guidance.

c. The entire focus of US assistance under FID is to assist an HN, if possible, in anticipating, precluding, and as a last resort, countering an internal threat. The type of planning necessary is dictated by the type or types of support being provided. **Support in anticipating and precluding threats is preventive in nature and is likely to require a mix of indirect support and direct support not involving combat operations.** An existing threat is likely to require responses that span all categories of FID support, to include US combat operations. A detailed discussion of employment considerations is included in Chapter VI, "Foreign Internal Defense Operations."

d. Prior to a training team or JTF conducting a FID mission in a foreign country, many levels of policy development and operation planning will take place. The specific mission to be conducted can range from participating in a multinational exercise to training an HN force on basic infantry skills. FID missions will fall under two major categories—those under the responsibility of DOD and those under DOS. To a training team in the HN, the category may seem irrelevant; however, the activity or program a training team has been deployed to participate in is governed by specific rules, funding, and conditions, depending on if the program falls under DOD or DOS oversight. The majority of DOD and DOS activities are incorporated into the theater planning process. Through the theater planning process, identified activities are intended to help shape the theater in which the activities will be conducted. Depending on whether the mission has originated through DOD or DOS, how, where, and at what level the planning, coordination, and resourcing takes place will vary. For example, Title 22, USC, governs DOS programs and indicates participants in these programs are noncombatants. Programs under Title 10, USC, authorities do not restrict participants from being combatants.

2. Planning Imperatives

FID has certain aspects that make planning for it complex. Some basic imperatives when integrating FID into strategies and plans are:

a. **Maintain HN Sovereignty and Build Legitimacy.** If US military efforts in support of FID do anything to undermine the sovereignty or legitimacy of the HN government, then they have effectively sabotaged the FID efforts. Ultimately, **FID operations are only as successful as the HN's IDAD program.**

b. **Understand long-term or strategic implications and sustainability of all US assistance efforts** before FID operations commence. This is especially important in building HN development and defense self-sufficiency, both of which may require large investments of time and materiel. Comprehensive understanding and planning will include assessing the following:

(1) The end state for development.

(2) Sustainability of development programs and defense improvements.

(3) Acceptability of development models across the range of HN society, and the impact of development programs on the distribution of resources within the HN.

(4) Second-order and third-order effects of socioeconomic change.

(5) The relationship between improved military forces and existing regional, ethnic, and religious cleavages in society.

(6) The impact of improved military forces on the regional balance of power.

(7) Personnel life-cycle management of military personnel who receive additional training.

(8) The impact of military development and operations on civil-military relations in the HN.

c. **Tailor military support to FID for the operational environment and the specific needs of the supported HN.** Choices in equipment and training conducted during FID operations may affect future interoperability capabilities. Consider the threat as well as local religious, social, economic, and political factors when developing the military plans to support FID. Overcoming the tendency to use a US frame of reference is important because this potentially damaging viewpoint can result in equipment, training, and infrastructure not at all suitable for the nation receiving US assistance.

d. **Ensure Unity of Effort / Unity of Purpose.** As a tool of US foreign policy, FID is a national-level program effort that involves numerous USG agencies that may play a dominant role in providing the content of FID plans. Planning considers and, where appropriate, integrates all instruments of national power and IGOs, NGOs, and HN capabilities in order to reduce inefficiencies and enhance strategy in support of FID and HN IDAD efforts. An interagency plan that provides a means for achieving unity of effort among USG agencies is described in Appendix C, "Illustrative Interagency Plan for Foreign Internal Defense."

e. **Understand US Foreign Policy.** NSC directives, plans, or policies are the guiding documents. If those plans are absent, JFCs and their staffs must find other means to understand US foreign policy objectives for an HN and its relation to other foreign policy objectives. They should also bear in mind that these relations are dynamic, and that US policy may change as a result of developments in the HN or broader political changes in either country. DOD planners should seek guidance from the COM and country team in interpreting foreign policy and guiding US efforts in a particular country.

f. **Understand the Information Environment**

(1) With the advent of instant or nearly instant communications and media access in even the most remote regions, US FID efforts in any HN may be scrutinized more closely within the region, surrounding regions, or even globally. In addition, FID operations may affect countries throughout the region or even cause international debate and opposition. In some theaters, traditional rivalries and hostility toward the US will be a factor. For example, US assistance to a nation with long-standing adversaries in the area may be perceived by those adversaries as upsetting the regional balance of power. Although it is increasingly an untenable position, some nations within a region or elsewhere internationally may consider the HN to be within its "sphere of influence." The ethnic rivalries in the Balkans and the quasi-religious dogmas of the jihadists promulgating the war on terrorism show that history, even a millennium old, can still foster fanatic resistance to US FID efforts. These examples highlight the propensity of some internal threats to use revisionist history in their propaganda.

(2) Proactive PA programs can accurately depict US efforts. Adversary propaganda requires a concerted US PSYOP program to defeat it. The US can pursue a proactive PSYOP effort in the HN and neighboring regional countries to prepare key TAs for US FID operations. In addition, PSYOP can exploit early successes in the HN. US commanders must consider friendly, neutral, and hostile nations in the supported HN region and envision how they will perceive US support. PSYOP and PA can be coordinated to address regional, transregional, and if applicable, global audiences that may have (or perceive they have) a stake in any US FID operations.

(3) Planners must be mindful that all FID operations have the potential to move quickly from obscurity to the center stage of global media. Properly trained, culturally and politically attuned forces are necessary for successful mission accomplishment.

g. **Sustain the Effort.** This includes planning for both the US sustainment effort and also the efforts necessary for the HN to sustain its operations after the US or multinational forces depart.

3. **Department of Defense Guidance**

a. **Military planning for FID is initiated at the combatant command level.** GCCs base strategy and military planning to support FID on the broad guidance and missions provided in the JSPS. This section will briefly discuss the major JSPS guidance documents and their relation to the CCDR's FID planning process. Only those documents most relevant to the FID planning process will be discussed.

For further information on joint planning, refer to JP 5-0, Joint Operation Planning, *and Chairman of the Joint Chiefs of Staff Instruction (CJCSI) 3100.01B,* Joint Strategic Planning System.

b. The NMS supports the aims of the NSS and implements the National Defense Strategy (NDS). The NMS conveys the CJCS message to the joint force on the strategic direction the Armed Forces of the United States should follow to support the NSS and NDS. It describes the ways and means to **protect** the US, **prevent** conflict and surprise attack, and **prevail** against adversaries who threaten our homeland, deployed forces, allies, and friends.

c. **The Joint Strategic Capabilities Plan.** The JSCP implements the strategic policy direction provided in the GEF and initiates the planning process for the development of campaign, contingency, and posture plans. The GEF integrates DOD planning guidance into a single, overarching document and contains DOD guidance previously promulgated by several documents. **Through the guidance and resources provided in the JSCP, the GCCs develop their operation plan (OPLAN) and operation plan in concept format to support FID.** Generally, the JSCP provides guidance important to FID in the following areas.

(1) The JSCP provides general taskings to the GCCs that may mandate military support to a FID operation or provide the strategic guidance and direction from which GCCs may deduce military missions to support FID.

(2) The JSCP supplemental instruction provides **additional planning guidance, capabilities, and amplification of tasking for planning in specified functional areas.** The supplemental instruction impacts the military planning and execution to support FID; however, four functional areas are directly tied to FID, as described in Figure IV-1.

d. *Guidance for Employment of the Force.* The GEF provides the foundation for all DOD interactions with foreign defense establishments, and supports the President's NSS. With respect to SC, this guidance provides direction with respect to IW, building partnership capacity, and stability, security, transition, and reconstruction. SC tools are

Figure IV-1. Foreign Internal Defense Operations Functional Areas Associated with the Joint Strategic Capabilities Plan

discussed in Chapter I, "Introduction," paragraph 6, "Department of Defense Foreign Internal Defense Tools."

e. The GCC, using an integrated priority list, also identifies requirements to support FID efforts and request authorization and resourcing.

4. General Theater Planning Requirements

a. GCCs may develop theater strategies and campaign plans that support taskings by the CJCS in the JSCP. Regardless of how commanders may tailor the planning process, **military activities in support of FID requirements are integrated into concepts and plans from the strategic level down to the tactical level.**

b. **Theater strategy** translates national and alliance strategic tasks and direction into long-term, regionally focused operational tasks and direction to accomplish specific missions and objectives. The NMS, GEF, and JSCP guide the development of this strategy that incorporates peacetime and war objectives and reflects national and DOD policy and guidance. **Peacetime goals will normally focus on deterring hostilities and enhancing stability in the theater. FID is an integral part of this strategy.** The determination of the desired end state for the theater is an important element in the strategy process. This determination establishes the theater's strategic direction on which commanders and their staffs base campaign plans as well as other plans. In general, the theater strategy will normally include an analysis of US national policy and interests, a strategic assessment of the AOR, a threat analysis, the commander's vision, and a statement of theater missions and objectives.

c. **TCPs are operational extensions of the theater strategy.** They provide the commander's vision and intent through broad operational concepts and provide the framework for supporting OPLANs.

For further information on campaign plans, refer to JP 3-0, Joint Operations, *and JP 5-0,* Joint Operation Planning.

d. **Security Cooperation Planning.** SC is the means by which DOD interacts with foreign defense establishments to build defense relationships that promote specific US security interests, develop allied and friendly military capabilities for self-defense and multinational operations, and provide US forces with peacetime and contingency access to an HN. The GCC's TCP is the primary document that focuses on the command's steady-state activities, which include operations, SC, and other activities designed to achieve theater strategic end states. Direction for the GCC is provided through the GEF and the JSCP. This guidance provides regional focus and SC priorities. Services develop campaign support plans that focus on activities conducted to support the execution of their GCC's campaign plan, and on their own SC activities that directly contribute to the campaign end states and/or DOD component programs in support of broader Title 10, USC, responsibilities.

For additional information on SC planning, refer to JP 5-0, Joint Operation Planning, *and the* Guidance for Employment of the Force.

e. The Joint Operation Planning and Execution System (JOPES) is the approved system for conventional operation planning and execution. The JOPES contingency planning process interrelates with the NMS and other planning documents to develop OPLANs. **The contingency planning process is particularly applicable to FID planning, since most military activities in support of FID should be planned well in advance as part of a larger strategy or campaign.**

5. Planning Procedures and Considerations

a. JOPES provides the framework for planning functions that form the basis for both contingency and crisis planning. Figure IV-2 depicts this framework and the relationships between operational activities, planning functions, and products. This is fully described in both JP 5-0, *Joint Operation Planning,* and Chairman of the Joint Chiefs of Staff Manual (CJCSM) 3122.01A, *JOPES Volume I, Planning Policies and Procedures.* FID planning follows these guidelines and is further described below.

b. **Strategic Guidance.** There are **three methods by which the GCC identifies requirements for military activities to support FID.** Plans for these specific military operations in support of FID become part of the overall theater effort and are incorporated into all levels of planning.

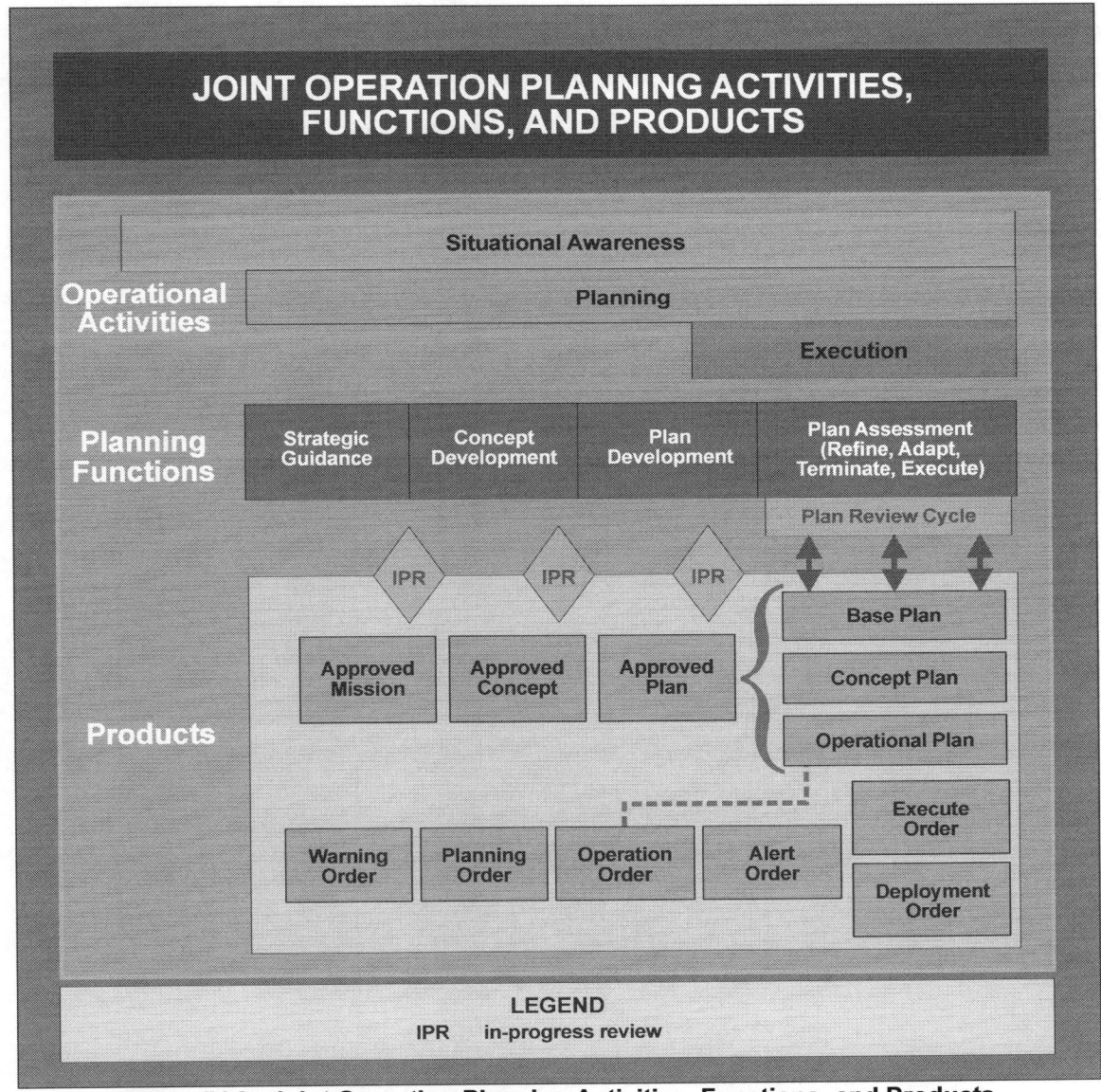

Figure IV-2. Joint Operation Planning Activities, Functions, and Products

(1) **Top down through the JSPS.**

(2) **Bottom up from an HN or country team in the GCC's AOR.** These include both SA and other SC requests. SA requests are forwarded by the SC officer on the country team directly to the implementing agency. The GCC is informed of such requests but execution is outside the JSPS. For other SC requests the GCC may forward those requests to SecDef for authorization. The GCC may authorize these support missions whenever they are in accordance with US law and directed through the JSPS.

(3) **GCC Initiated.** Military support to FID efforts that is not directed under an existing specified or implied mission may be identified. The GCC endorses these requirements and obtains authorization from SecDef.

c. **Concept Development.** Figure IV-3 provides an overview of the major concept development stages. FID considerations include the following.

(1) **Mission Analysis.** Before beginning FID planning, **the commander's staff will conduct a thorough mission analysis of the operational environment and threat.** This mission analysis establishes the operational framework for FID concept

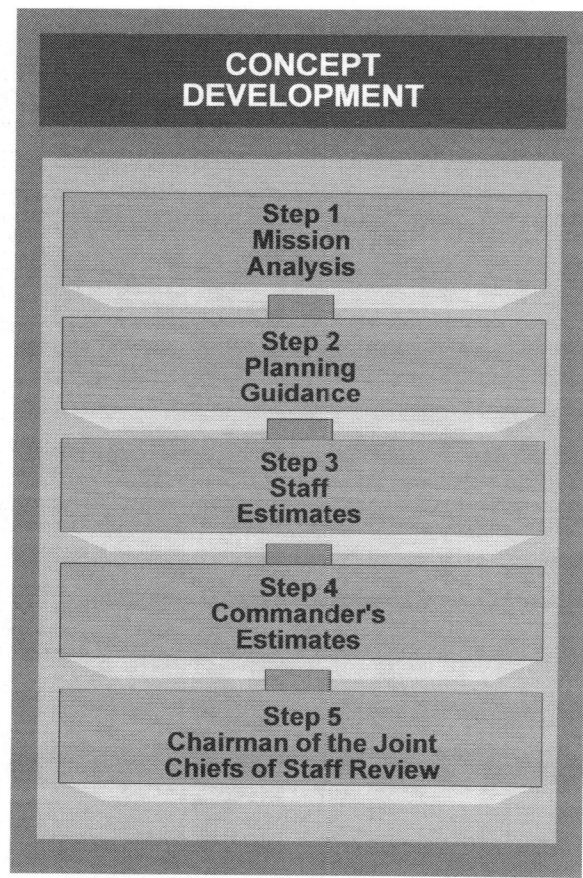

Figure IV-3. Concept Development

development and planning. The following areas are considered when developing the concept.

(a) **Threats to HN IDAD.** Threats may be specific, such as illicit drugs or terrorism, or they may be more general as in social unrest and instability. **Identification of the root cause is key** so that military activities in the FID plans may target long-term causes rather than short-term symptoms. Appendix B, "Joint Intelligence Preparation of the Operational Environment to Support Foreign Internal Defense," provides detailed guidance for conducting intelligence preparation of the operational area necessary for effective FID planning.

(b) **The HN Social, Economic, and Political Environment.** FID efforts are intended to support IDAD programs in a manner that is acceptable to the HN's cultural and political realities. **The capability of the HN government and leadership as well as existing treaties and social infrastructure are all factors that planners must consider.** This step may result in the conclusion that the best solution from the US perspective may not be the best solution for the supported HN. This proposed solution may be outside the realm of FID and may be better accomplished through other means. This situation must be resolved in diplomatic channels between the USG and the HN. For example, a treaty may meet US goals and objectives independent of the HN IDAD program and interests.

(c) Mission analysis for military operations in support of FID normally will be conducted in conjunction with normal operation planning. During this phase, commanders and their staffs must analyze the assigned tasks, develop a mission statement, formulate subordinate tasks, and prepare guidance for the commander's approval. The first two of these are discussed below.

<u>1</u>. **Analyze Assigned Tasks.** Tasks to support FID will be specified and implied and may range from supporting SA efforts to providing forces to conduct FHA efforts.

<u>2</u>. **Develop a Mission Statement.** The mission statement guides much of the remaining planning process. The mission statement will provide the who, what, when, where, why, and possibly how. The theater mission established by the GCC may be general, but could also identify FID-related tasks. The mission statement focuses on the priority threats to the security and stability of the HN.

(2) **Planning Guidance.** During planning guidance, the commander outlines tentative COAs, additional assumptions, and a planning directive to the staff and subordinate commanders. **Several important FID guidelines must be considered** to ensure that future planning results in the most efficient employment of the commander's resources.

(a) **Legal Authorizations and Restrictions.** The legal restrictions governing military activities in support of FID are complex and subject to changing US legislation. **The staff legal advisor has an active role in the FID planning process.** The basic funding authorizations for military activities in support of FID come through either the FAA, AECA, or through DOD operation and maintenance (O&M) funding sources. **While GCCs may use O&M funding for specified and limited FID activities such as HCA, they may not use O&M funds for SA or exceed strict dollar limits on military construction projects.** Appendix A, "Legal Considerations," provides a look at the legal aspects of FID.

(b) **Third Country Interests. US FID efforts may impact countries throughout the region.** US FID activities may impact conditions and perceptions throughout the region, particularly in neighboring countries. These factors will not dictate US policy, but require careful evaluation and consideration and may impact the design of specific operations. Commanders should use active information programs to accurately depict US efforts to influence perceptions and to defeat adversary propaganda efforts.

(c) **Restrictive Use of Force.** US combat operations in FID will generally require a judicious selection and employment of forces. The purpose of such selection and employment is to ensure that the HN military and other civilian security forces rapidly accept the responsibility for its security and to minimize HN civilian casualties. Employment of nonlethal capabilities significantly enhances options for force application and provides a means to achieve scalable levels of force.

1. Specific ROE are developed and approved to support FID operations. They will normally differ from country to country, may differ within regions of the same country, and are likely to be more restrictive in FID than in other operations. However, the existence of such restrictive ROE does not preclude the US from employing a proportional level of force or using the force required to protect the lives of deployed US personnel. It also will not restrict the authority of on-scene commanders to respond with appropriate force and use all necessary means available and take all appropriate actions in self-defense when they perceive a threat to the safety and security of US personnel.

2. Standing rules of engagement (SROE) are provided for US forces as stand-alone guidance that can be tailored to meet mission-specific requirements. SROE apply in the absence of specific ROE from higher authority in the form of supplemental measures; these supplemental measures may be provided by, or requested from, higher authority to tailor ROE for a particular mission. GCCs may also augment the SROE in order to respond to mission and threat in their AOR. SROE do not apply to DOD civil support missions (i.e., support to civil authorities and civilian law enforcement agencies); routine Military Department functions, including FP duties, within US territory and territorial seas; homeland defense land missions within US territory; law enforcement and security duties on DOD installations (and off-installation, while conducting official DOD security functions), within or outside US territory; USCG units (and other units under their OPCON) conducting law enforcement operations; and US forces assigned to the

OPCON or tactical control of a multinational force (the ROE of the multinational force will be followed, if authorized by SecDef).

For additional information on SROE, refer to CJCSI 3121.01B, Standing Rules of Engagement/Standing Rules for the Use of Force for US Forces.

(3) **Staff Estimates.** The staff analyzes and refines tentative COAs during the staff estimate process of concept development. These detailed options serve as the foundation for the commander's decision to select a COA. Military options to support FID under consideration can involve any of the categories of indirect support, direct support (not involving combat operations), or combat operations. **The following three estimates have specific implications for the development of an effective FID plan.**

(a) **The intelligence estimate is essential to accurately identify the threat upon which to base FID efforts.** The intelligence estimate supporting FID operations will have an orientation quite different from that of a conventional estimate. A comprehensive and intimate knowledge of the operational environment is essential in building this estimate. The intelligence estimate concentrates on adversary situation; adversary capabilities, to include their capacity to produce chemical, biological, radiological, and nuclear (CBRN) weapons and the location of CBRN materials, including toxic industrial materials, that could be used to produce such weapons or makeshift devices; an analysis of those capabilities; and finally, conclusions drawn from that analysis. **In FID, however, analysis must focus more on the local population and its probable reactions to potential US or opposition actions.** This requires knowledge of the ethnic, racial, economic, scientific, technical, religious, and linguistic groups in the HN, as well as their locations and an understanding of how they may perceive future operations. Understanding the operational environment and the HN's social, economic, and political systems is essential in order to effectively and efficiently support the HN's IDAD program. Appendix B, "Joint Intelligence Preparation of the Operational Environment to Support Foreign Internal Defense," discusses the FID-specific aspects of intelligence preparation that must be considered in order to successfully plan and execute FID operations.

(b) **The CMO estimate examines each military COA and how best CMO may influence the various COAs the commander is considering.** The focus of the CMO estimate development is on situation assessment rather than COA development. The purpose is to assemble information underlying a CMO concept of support that can be modified to support the overall concept of operations. CA will also complete area studies where operations are likely. For military operations to support FID, these assessments focus on social, economic, and political factors that relate to existing or potential lawlessness, subversion, or insurgency. These assessments may include overlays that show local demographics, civil supply support, public utilities, and population displacement. The CMO estimate shows how CMO can best be integrated into the overall operation and supports decisionmaking throughout an operation. CA support, like PSYOP, should be incorporated into all FID operations.

(c) **The PSYOP staff estimate process examines the potential impact of proposed US military operations.** Internal stability is closely connected to the HN populace's perceptions; thus their perceptions should be continually assessed and PSYOP adjusted accordingly. More specific guidance is included in the PSYOP annex to the appropriate plan. **At the combatant command level, PSYOP concepts and plans must be coordinated through the country team.** This coordination is discussed in more detail in Chapter VI, "Foreign Internal Defense Operations."

(d) In addition to the planning imperatives previously discussed, there are **several important FID guidelines** to consider when developing possible COAs. These guidelines are:

<u>1</u>. **Maximize Intelligence and CI Capabilities.** Identify political, economic, scientific, technical, and social threats, in addition to the conventional hostile military factions. This is a complex task, especially when working in an unfamiliar culture in which US personnel may have little or no experience and in which high-technology collection and processing equipment may be of little use. Despite this challenge, commanders and their staffs must integrate all available assets and use culturally trained specialists to define these threats and to appropriately tailor the COAs.

<u>2</u>. **Force Protection.** Commanders and staffs must evaluate operations closely in order to determine the increased risks of large deployments of US personnel in the area. In a high-threat condition, it may be prudent to delay the FID mission or commit a smaller element, rather than to commit a larger force that has a higher profile and is more difficult to protect. Depending on the mission, the JFC may consider seabasing some of the force as an alternative to reduce the footprint ashore.

<u>3</u>. **Account for Sustainment.** Commanders and staff must be cognizant of HN culture and their appreciation for sustainment, maintenance, and budgeting. IDAD programs are frequently at risk in the out years because they may not receive adequate HN investment. This applies both to physical infrastructure and programs to improve human capital, good governance, and other intangibles.

<u>4</u>. **Measures of Effectiveness. Clearly define and focus on MOEs.** The success of US FID activities can be measured only in terms of the success of the HN's IDAD program. MOEs should focus on long-term, attainable objectives rather than short-term targets, limited objectives, or over-ambitious development goals. In addition, establish transition points that define when the supported HN will, incrementally, assume responsibility for the total IDAD effort.

d. **Plan Development.** Plan development (Figure IV-2) begins after the GCC's strategic concept is fully developed. **This phase matches mission requirements against available resources. This action is particularly important because a large portion of the force needed to conduct FID is in the Reserve Component (RC) and, in most cases, is unavailable (short of use of Presidential Reserve Call-up Authority) for long-term operations.** In major military operations in support of a FID operation, the commander must consider availability of forces (both Active Component [AC] and RC) to support the mission requirements. Use of RC forces, including USCG Reserve, may alleviate shortfalls and assist organizing and tailoring resources to carry out the military support to FID as efficiently as possible. Similarly, leveraging the mutually supporting capabilities of the US Navy (USN) and USCG can serve as a force multiplier for FID activities in the maritime domain. HN, third-party nations, and USG interagency coordination (country team) remains paramount during plan development. This is an extension of the coordination that began during the plan initiation phase.

e. **Plan Assessment.** Commanders and their staffs should consider that many FID objectives will involve a long-term effort, and that MOEs may be difficult to evaluate in the short term.

f. **Supporting Plans. Supporting FID plans may come from a variety of units such as Service or functional component commanders, subordinate unified commanders, JTF commanders, supporting CCDRs or subordinate CA, PSYOP, engineer, medical, transportation, special forces (SF) or other combat units.** Supporting plans cover mobilization, deployment, and employment, and are the "how to" of the plan.

g. **Other Planning Considerations**

(1) For personnel recovery operations that may occur during FID, refer to JP 3-50, *Personnel Recovery*.

(2) A key component of developing the training plan will be an agreement (memorandum of agreement or letter of agreement) between the HN and the joint force conducting the training. These are, in effect, signed contracts detailing the specifics of what is being provided by each party and should be carefully reviewed by the legal advisor for any instrument that is potentially an international agreement. Training plans at the operational level will vary based on HN needs and unit training capabilities. An assessment for the training to be conducted should begin with a pre-mission site survey by all joint force units deploying in the HN.

6. **Planning for Force Protection**

a. **JFCs and their subordinate commanders must address FP during all phases of FID operations,** from planning through deployment, employment, and redeployment. All aspects of FP must be considered and threats minimized to ensure maximum operational success. JFCs and their subordinate commanders must implement FP measures appropriate to all anticipated threats, to include terrorists and the use of CBRN material as a weapon.

b. Supported and supporting commanders must ensure that deploying forces receive thorough briefings concerning the threat and personnel protection requirements prior to and upon arrival in the operational area.

c. In addition, JFCs and their subordinate commanders must evaluate the deployment of forces and each COA for the impact of terrorist organizations supporting the threat and those not directly supporting the threat but seeking to take advantage of the situation within the HN.

d. JFCs also must plan for CI support to the FID operation and should incorporate umbrella concepts for CI FP source operations in the planning process.

e. For the foreseeable future, the threat of terrorism is a constant factor, even in those nations with advanced domestic security infrastructures. FID operations require proactive FP measures and consideration of the following:

(1) Street or organized crime.

(2) Foreign intelligence services.

(3) Local populace animosity or demonstrations.

(4) HN regional transportation systems.

(5) HN military equipment, training, and procedural deficiencies.

(6) WMD active defense and CBRN passive defense.

(7) Weather, terrain, and environment.

(8) Contaminants and pollution.

(9) Health threats and HN health care capabilities.

(10) Mines and unexploded ordnance.

(11) Political groups, parties, or individuals.

(12) Regional cross-border threats.

For further information on FP, refer to JP 3-0, Joint Operations, *and JP 3-07.2,* Antiterrorism. *For further information on CBRN, refer to JP 3-11,* Operations in Chemical, Biological, Radiological, and Nuclear (CBRN) Environments, *and JP 3-40,* Combating Weapons of Mass Destruction.

7. Foreign Internal Defense Assessment

a. Commanders and staffs must realize that planning for joint FID operations is an integrated process. It is totally integrated into theater planning and is reflected in planning documents extending from the JSCP and the GCC's strategy down to subordinate joint force and Service component supporting plans. During the concept development of FID planning, a broad approach to JIPOE must be considered. In addition, PSYOP and CMO estimates are included in the decision-making process to formulate adequate and feasible COAs from the FID standpoint; how best CMO can influence the operational concept. Finally, the staff must review FID plans for adequacy, acceptability, feasibility, and compliance with joint doctrine, and they must ensure the creation of appropriate supporting plans to support theater-level operations.

b. Primarily, those responsible for conducting FID operations assess what activities to conduct. DOS representatives work with foreign governments and DOD representatives work with foreign military personnel to develop IDAD programs that are consistent with US foreign policy objectives and useful to the country concerned. The theater SC planning process is used to assess currently implemented programs and exercises. Representatives assess the previous programs for relevancy and success in relation to country-specific objectives and to the overall goals within the region. Programs are assessed on the basis of key trends, shortfalls, future opportunities, and challenges.

8. United States Foreign Internal Defense Capabilities

a. **Military Resources.** SOF have traditionally been the forces of choice for FID due to their extensive language and cultural training and their regional focus. However, FID requirements in modern conflict zones far outsize the capability of limited SOF. CF can, and should, be prepared to conduct FID operations where appropriate. In fact, CF have a distinct advantage in certain FID operations where their HN counterparts have a similar conventional mission. Conventional commanders must assign the best qualified, mature, and culturally attuned Service members to training and advisory missions. Both CF and SOF participating in FID operations outside of their normal area of expertise will require augmentation from interpreters fluent in the HN language and cultural training.

(1) US Army

(a) Army FID operations are primarily aimed at developing and improving HN ground force and landpower capabilities through ground force advisor operations in coordination with SA programs. A principal US activity for conducting FID is the transfer of major items (weapon systems and related support items) to selected HNs, primarily through SA. The Army often facilitates such transfers through operational and strategic assessments, ground force studies, SA-funded equipment refurbishment, training on specific weapon systems and support items through SA-funded MTTs, and SA case management and oversight. Delivery of FMS items can be performed in conjunction with multinational operations and contingencies and with other training programs conducted by the GCCs and USG agencies. FID operations can establish a US presence, build rapport, achieve integration of forces, and build a foundation for future regional cooperation. If necessary, commanders can employ an even greater range of capabilities and resources in more direct forms of FID support when HN ground units are inadequately sized or structured to make necessary and timely contributions to their own defense effort. Principal initiatives to accomplish FID objectives include the following:

    <u>1.</u> Facilitate the transfer of US defense articles and services under the SA program to eligible foreign government aviation units engaged in IDAD operations.

    <u>2.</u> Assess foreign military ground force capabilities and provide direction or recommendations toward improving HN landpower employment and sustainment methods.

    <u>3.</u> Educate foreign military force senior officers and civilians in how to appropriately use military power.

    <u>4.</u> Train foreign military forces to operate and sustain indigenous ground resources and capabilities.

    <u>5.</u> Advise foreign military forces and governmental agencies on how to employ ground forces in specific operational situations.

    <u>6.</u> Assist foreign ground forces in executing specific missions or contingency operations. Assistance can take on many forms, but generally includes hands-on assistance in enabling capabilities such as intelligence operations, military police/detainee operations, CA/civil-military cooperation, sustainment/logistical support, and health services support.

    <u>7.</u> Facilitate force integration for multinational operations.

    <u>8.</u> Provide direct support to host countries by using Army resources to provide intelligence, communications capability, mobility, logistic support, and Army aviation support.

For further guidance and detail on US Army FID capabilities, refer to Field Manual (FM) 3-07, Stability Operations, *and FM 3-07.1,* Security Force Assistance.

    (b) **US Army Special Operations Forces.** Army special operations forces (ARSOF) may conduct FID operations unilaterally in the absence of any other military effort, support other ongoing military or civilian assistance efforts, or support the employment of CF. The ARSOF core task of FID is only one component of an HN's IDAD policy and programs. The strategic end state of FID is an HN capable of successfully integrating military force with other instruments of national power to eradicate internal threats. From an SOF perspective, FID efforts are successful if they preclude the need to deploy large numbers of US military personnel and equipment. FID, as part of NA, takes place in an arena that may include support to COIN, counterterrorism (CT), PO, DOD support to CD operations, and/or FHA. These activities may include FID operations as an integral component in supporting the fight against internal threats. Small teams may conduct FID training in remote areas in HNs where the

US force commitment is small. When this is the case, ARSOF units may be the primary or sole trainers of HN military or paramilitary personnel. In other cases, large numbers of forces and units may be involved in direct support (not involving combat operations). Training of HN military may continue while US forces are providing direct support and, to some degree, even during combat operations. A commander must balance the need to operationally employ ARSOF against their unique abilities (possibly unparalleled by the CF available) to train critical capabilities in the HN military.

For further guidance and detail on ARSOF FID capabilities, refer to FM 3-05.137, Army Special Operations Forces Foreign Internal Defense.

(2) US Marine Corps

(a) Marine Corps forward deployed forces provide flexible, protected, and self-sustained sourcing solutions to a wide range of SC missions. Marine Corps SC Marine air-ground task forces (MAGTFs) may be tasked with building HN security capacity and supporting partner nation security efforts in a specific regional area.

(b) **Marine Corps Special Operations Forces (MARSOF).** MARSOF is the USMC component to USSOCOM. MARSOF supports FID operations through the Marine special operations advisor group (MSOAG), which provides tailored military combat-skills training and advisor support for identified foreign forces in order to enhance their tactical capabilities and to prepare the environment as directed by USSOCOM. Marines and Sailors of the MSOAG train, advise, and assist friendly HN forces—including naval and maritime military and paramilitary forces—to enable them to support their governments' internal security and stability, to counter subversion and to reduce the risk of violence from internal and external threats. Marine Corps Special Operations Command (MARSOC) teams engage globally as directed by USSOCOM in support of TSOC engagement priorities and conduct FID with partnered foreign SO and CT forces. MSOAG deployments are coordinated by MARSOC, through USSOCOM.

(3) **US Navy.** Support to FID is a core competency for Navy CF. The Navy Expeditionary Combat Command is composed of several subordinate commands that are able to provide unique capabilities in support of FID. Navy military CA units work directly with the civil authorities and civilian populations within a maritime area of operations helping to enhance security and stability. Expeditionary training units deliver focused maritime capabilities training to HN personnel across a wide array of maritime competencies.

(a) Construction capabilities in support of operating forces, as well as civic projects for other nations, to include building roads, bunkers, airfields, and bridges, as well as providing responsive support for disaster recovery operations.

(b) Expeditionary support capable of conducting port and air cargo handling missions, customs inspections, contingency contracting capabilities, fuels distribution, freight terminal and warehouse operations, postal services, and ordnance reporting and handling.

(c) Expeditionary diving support capable of conducting salvage, search and recovery, harbor clearance, underwater cutting and welding, construction, inspection, and repair of ocean facilities including wharves, piers, underwater pipelines, moorings, boat ramps, and limited repairs on ships.

(d) Maritime expeditionary security forces capable of defending designated assets and maritime infrastructure in the near coast environment in support of SFA missions.

(e) Riverine forces capable of establishing and maintaining control of rivers and waterways for military and civil purposes, denying their use to hostile forces, and combating sea-based terrorism.

(f) Global Fleet Stations (GFSs) support HCA missions. GFSs are able to conduct shaping and stability operations in support of FID missions.

(g) US Navy Seals and Naval Small Craft Instruction and Technical Training School also perform FID operations as a core mission.

(4) **US Air Force**

(a) Air Force FID operations are primarily aimed at developing and improving HN airpower capabilities through air advisor operations in coordination with SA programs. A principal US instrument for conducting FID is the transfer of major items (weapon systems and related support items) to selected HNs, primarily through SA. The Air Force often facilitates such transfers through operational and strategic assessments, airpower studies, SA-funded aircraft refurbishment, airlift of SA-funded defense articles, training on specific weapon systems and support items through SA-funded air advisory and technical MTTs, and SA case management and oversight. Delivery of FMS items can be performed in conjunction with multinational operations and contingencies and with other training programs conducted by the GCC and USG agencies. Air Force FID operations can establish a US presence, build rapport, achieve integration of forces, and build a foundation for future regional cooperation. If necessary, commanders can employ an even greater range of capabilities and resources in more direct forms of FID support when HN aviation units are inadequately sized or structured to make necessary and timely contributions to their own defense effort. Principal Air Force initiatives to accomplish FID objectives include the following:

1. Facilitate the transfer of US defense articles and services under the SA program to eligible foreign government aviation units engaged in IDAD operations.

2. Assess foreign military aviation capabilities and provide direction or recommendations toward improving HN airpower employment and sustainment methods.

<u>3.</u> Train foreign military forces to operate and sustain indigenous airpower resources and capabilities.

<u>4.</u> Advise foreign military forces and governmental agencies on how to employ airpower in specific operational situations.

<u>5.</u> Assist foreign aviation forces in executing specific missions or contingency operations. Assistance can take on many forms, but generally includes hands-on assistance in combat support capabilities such as aircraft maintenance, fuels, health service support (HSS), and aviation medicine.

<u>6.</u> Facilitate force integration for multinational operations.

<u>7.</u> Provide direct support to host countries by using Air Force resources to provide intelligence, communications capability, mobility, logistic support, and airpower.

For further guidance and detail on Air Force FID capabilities, refer to Air Force Doctrine Document (AFDD) 2-3.1, Foreign Internal Defense.

(b) **Air Force Special Operations Forces (AFSOF).** AFSOF support FID operations by working by, with, and through HN aviation forces from the ministerial level to the tactical unit. When required, AFSOF provide persistent manned and unmanned intelligence, surveillance, and reconnaissance (ISR), mobility, and precision engagement. AFSOF maintain specially trained combat aviation advisors to assess, train, advise, and assist HN aviation capability thereby facilitating the availability, reliability, safety, and interoperability of these forces. Additionally, AFSOF special tactics teams enhance air to ground interface, synchronizing conventional and special operations during COIN operations.

(5) **US Coast Guard.** The USCG possesses broad authorities across the spectrum of military, law enforcement, regulatory, and intelligence activities in support of FID. The USCG serves as a model "maritime service" for the navies and coast guards of emerging democratic nations.

(a) Many of the world's navies and coast guards have a mix of military, law enforcement, resource protection, and humanitarian functions very similar to those of the USCG. A common constabulary and multi-mission nature promotes instant understanding and interoperability and makes USCG a valued partner for many naval and maritime forces. The USCG has a long history of providing training and support to maritime forces around the world.

(b) Building effective maritime governance requires engagement beyond navies and coast guards. It requires integrated efforts across agencies and ministries, as well as private sector commitment. The USCG has this expertise by virtue of its broad statutory missions, authorities, and civil responsibilities; membership in the intelligence community; and strong partnerships with industry. USCG routinely engages other nations through multiple ministries and can offer a model maritime code that countries can use to improve their laws and regulations.

(c) The USCG is the lead agency for maritime CD interdiction, and it is the principal maritime law enforcement agency of the US. In addition, subject to international agreements, the USCG may patrol or conduct pursuit, entry, and boarding operations in the territorial waters of other countries.

(6) The National Guard's State Partnership Program establishes partnerships between foreign countries and participating US states sponsoring and contributing through reachback to support DOD's SC programs.

b. **Interagency Resources.** JFCs consider capabilities and availabilities of interagency resources when planning for FID operations. During execution, coordination and synchronization of military operations with interagency initiatives and activities is normally coordinated by the country team. This allows for the most capable and similar organizations to work together in developing effective and efficient HN resources. For example, DOJ and DOS can send law enforcement specialists overseas to train and advise HN police forces; police are best trained by other police. The nature of the operational environment and legal considerations are also important. USCG theater SC activities reach beyond normal military-to-military relations to a broader HN maritime audience, including, but not limited to, law enforcement agencies, maritime administrations, and transport ministries. The USCG is uniquely positioned to engage through traditional and non-traditional defense, law enforcement, and maritime safety operations.

c. **Contractor Support.** In some cases, additional training support from contractors enables commanders to use joint forces more efficiently. For example, contracted police development capabilities through the DOS's INL can provide expertise not resident in the uniformed military. Contractor support can provide HN training and education, including the following:

(1) Institutional training.

(2) Developing security ministries and headquarters.

(3) Establishing administrative and logistic systems.

CHAPTER V
TRAINING

> *"A government is the murderer of its citizens which sends them to the field uninformed and untaught, where they are to meet men of the same age and strength, mechanized by education and discipline for battle."*
>
> **Light-Horse Harry Lee III, 1756-1818**

SECTION A. TRAINING JOINT FORCES FOR FOREIGN INTERNAL DEFENSE

1. General

a. A tenet of our national military strategy is to assist allies, coalition partners, and the governments of threatened states in resisting aggression. This strategic imperative demands that joint forces strengthen their abilities to assist, train, and advise foreign military and security forces.

b. FID may be conducted by a single individual in remote isolated areas, small groups, or large units involved in direct support (not involving combat operations) or combat operations. In many of these situations, conventional US forces will be operating in unfamiliar circumstances and cultural surroundings.

c. FID operations may be conducted in uncertain and hostile environments. Combined with the stresses of operating in a foreign culture, this may require training that is not routinely offered to CF.

d. **Commander, United States Special Operations Command** (CDRUSSOCOM) is charged by legislation with training assigned forces to meet mission taskings (including FID) and to ensure their interoperability with CF as well as other SOF. Continuing individual education and/or professional training peculiar to SO are the responsibility of CDRUSSOCOM.

For further information on CDRUSSOCOM training and education responsibilities, refer to JP 3-05, Joint Special Operations.

2. Training and Skills Needed for Success in Foreign Internal Defense

The following subparagraphs highlight some of the training needed for successful military operations in support of FID.

a. **Overall US and Theater Goals for FID.** Personnel engaged in FID operations must understand the overall goals and objectives of the supported GCC. This knowledge is similar to understanding the commander's intent in conventional operations. An understanding of these goals provides a framework to determine if his or her actions and operations support overall theater objectives.

b. **Area and Cultural Orientation.** Knowledge of the operational area is required to maximize the effectiveness of military operations in support of IDAD programs. It is difficult to successfully interact with the HN if individuals conducting FID do not have an understanding of the background of the nation, the culture, and the customs.

c. **Language Training.** It is very important for all personnel conducting FID operations to be able to communicate with HN personnel in their native language. Language capabilities can significantly aid trainers and others who have daily contact with HN military personnel and the local population. Personnel can function much more effectively if they receive language training in the target language prior to deployment. Even if they do not have extensive familiarity with the language, their use of basic expressions can help establish rapport with HN counterparts. Although language training is important, it is equally important for personnel conducting FID operations to conduct training on working with and speaking through interpreters.

d. **Standards of Conduct.** It is extremely important that all US military personnel understand the importance of the image they project to the HN population. This impression may have a significant impact on the ability of the US to gain support for FID operations or the HN to gain long-term support for its IDAD program. US standards of conduct are, to a great extent, an example of the professionalism and respect inherent in our system—a respect and professionalism that can be passed by example to the HN military and police forces and population. Refresher training in standards of conduct, which complement cultural awareness training, should be mandatory for all personnel involved in FID operations.

e. **Information Collection.** Because of their proximity and access to the local populace, personnel conducting activities in support of FID are passive information collectors and have valuable information for the intelligence system. These units and teams absorb information on the social, economic, and political situation that is essential to the operational area evaluation (OAE) discussed in Appendix B, "Joint Intelligence Preparation of the Operational Environment to Support Foreign Internal Defense." This information would be very difficult for more distant collectors to obtain. Care must be taken, however, to ensure that the relationship with their HN counterparts is not damaged by these activities. Personnel involved in FID operations must know and understand their responsibilities in these areas.

f. **Coordinating Relationships with Other USG Agencies, NGOs, and IGOs.** FID operations are likely to interact at all levels with other USG agencies, in addition to NGOs and IGOs. For example, SF, PSYOP, and CA elements may coordinate with the embassy PAO or cultural attaché, and CA may work closely with USAID. This type of coordination may be new to some military personnel; therefore, specific training or procedures may be required.

g. **Legal Guidelines.** In order to function effectively, personnel supporting FID activities must be aware of a variety of legal guidelines. These include provisions of applicable status-of-forces agreements (SOFAs) as well as restrictions on the transfer of equipment and on other types of assistance that may be provided. Because many military activities take place within the HN, applicable legal guidelines may include those of the HN government and the status of US personnel while in country (for example, existing SOFAs). Accordingly, a country law briefing, cultural orientation, and review of any international agreements affecting status-of-forces should be included in training.

Appendix A, "Legal Considerations," provides a look at the legal aspects of FID.

h. **ROE.** A thorough understanding of the ROE is very important to units involved in combat operations and for individuals involved in any military activities in support of FID.

i. **Tactical Force Protection Training.** FID activities often require small US elements to deploy in isolated areas to support threatened HN governments. This requirement makes for a potentially dangerous situation for US personnel. US forces must be prepared for these situations, with proper training in self-protection programs and measures. Training should include individual and collective techniques.

3. **Foreign Internal Defense Training Strategy**

a. Training to prepare for military operations to support FID requires that a broad range of areas be covered. The training also must be designed to support a mix of personnel, ranging from language-trained and culturally focused SOF to those untrained in the specific area where the FID operation is located. Some training, such as language qualification, requires an investment in time and money that will not be practical for all personnel. **A combination of institutional and unit-conducted individual and collective training will be required.**

b. **Institutional Training.** SOF receive extensive institutional training in language, cultural considerations, and instructional techniques as qualifications in their basic specialty. When available and in sufficient numbers, these personnel should be extensively used to train HN forces and facilitate liaison with the HN. CF that are tasked to provide training and serve as advisors and MTTs and to conduct joint and multinational operations with HN forces require language, cultural, and other training to prepare them for these missions. Some institutional courses are available that can be used by commanders in order to train personnel for FID missions. Listed below are some of the types of institutional training that is provided by one or more of the Services. Consult appropriate training catalogs for DOD course listings.

(1) Language training.

(2) Cultural awareness and interpersonal communications training.

(3) General FID and IDAD principles training.

(4) FP and AT awareness training.

(5) SA team orientation training.

(6) SA technical training.

c. **Unit Training.** Much of the training necessary to prepare personnel to support FID activities may be conducted within the unit. This training can be individually focused or, in the case of unit-size participation, may involve large-scale collective training. Training resources may be drawn from a variety of sources, but SOF are particularly valuable because of their area orientation and FID focus. When feasible, units should conduct operational rehearsals of the FID mission. These rehearsals allow participants to become familiar with the operation and to visualize the plan. Such rehearsals should replicate, as much as possible, the potential situations that a unit may encounter during a FID mission.

SECTION B. TRAINING HOST NATION FORCES

4. Training Plan

a. In addition to training for the joint force preparing to conduct FID, FID operations include training of HN security forces to build the capacity to support the IDAD strategy. The JFC develops a training plan based on a thorough mission analysis and assessment of the IDAD strategy, HN capabilities and needs, and the operational environment. This plan should be developed in conjunction with both the country team and with commanders of HN forces to ensure that comprehensive objectives are detailed.

b. **Assessment and Site Survey.** The first step in developing the HN training plan is the conduct of a site survey. In addition to identifying logistics requirements for trainers, the site survey must include an assessment of HN capabilities, drawing conclusions about gaps between capabilities and needs as identified in the IDAD strategy. The training assessment should, as a minimum, consider:

(1) HN doctrine and training literature, including differences from US doctrine;

(2) Constraints in HN resources and funding;

(3) Societal and military culture;

(4) Current level of HN proficiency;

(5) HN's ability (or inability) to field systems or equipment;

(6) Potential training facilities and areas based on projected training (e.g., ranges, urban terrain training sites);

(7) Proficiency of HN trainers;

(8) Equipment availability (e.g., radios, weapons, vehicles);

(9) C2 systems and procedures;

(10) Logistics systems and procedures;

(11) Cooperation level with US and HN intelligence agencies during operations and training exercises;

(12) Constraints on US IO and past HN IO miscues and history;

(13) Background and analysis of the main interagency actors in the region; and

(14) FP assessment.

c. **Training Plans**

(1) Upon completion of the assessment and site survey, the JFC should develop, with the HN leadership, a training plan for HN security forces. Because training all echelons of forces helps to achieve synchronized execution of mission-essential tasks throughout the HN force, HN training strategies must include multi-echelon training whenever possible. The training plan should include, as a minimum:

(a) List of agreed upon training objectives at each echelon of HN forces,

(b) Identification of units, commands, and leadership personnel requiring training across each echelon,

(c) Identification of FID resources required and how they will be provided, and

(d) An instrument (such as a memorandum or letter of agreement) between the HN and the JFC identifying what will be provided by each party.

(2) Individual units will develop plans for execution of the JFC's training plan.

5. **Training and Advising**

a. There is no distinct boundary between training assistance and advisory assistance. In general, **training assistance** is typically nonoperational in areas and under conditions where joint force personnel are not likely to be forced to engage any armed internal threat. In equal generality, **advisory assistance** may entail some operational advice and assistance beyond training in less secure areas. Joint force personnel render advisory assistance in relatively low-risk situations. However, the current global situation proves that any member of a joint force is, to some degree, in harm's way. Advisory assistance (and to a lesser degree training assistance) may involve situations that require personnel to defend themselves. Therefore, the commander and the US embassy accept greater risk. The difficulty in putting exact and unqualified definitions on either type of assistance is that both may take place through the entire range from indirect support to direct support (not involving combat operations) during the same FID operation. As long as risk is clearly defined, planned for, mitigated where possible, and, most importantly, deemed worth the potential cost, this lack of definition causes no inherent problems. Mitigation, however, may translate into exclusion of trainers or advisors from certain areas or dictate what type of force is deployed (i.e., SOF or CF with robust organic security capability).

b. Joint force elements typically develop, establish, and operate centralized training programs for the supported HN force. The joint force can also conduct individual, leader, and collective training programs for specific HN units. Training subjects run the gamut of military tasks, and training focus ranges from individual instruction through leader development to specialized collective training. The joint force can provide both training and advisory assistance in two ways. In either case, assistance may be provided under the chief, SCO in his role as the SDO in-country, other designated embassy official, or the TSOC, or JTF, depending on the C2 arrangement.

(1) Small teams may provide training or give operational advice and assistance to HN civilian, military, or paramilitary organizations.

(2) Individual personnel may be assigned or attached to the SCO to perform training and advisory assistance duties on a temporary or permanent basis.

c. **Training Assistance.** The agreement negotiated between US and HN officials provides the framework for the who, what, when, where, how, and why of military training assistance. Often, US doctrine, as prescribed in applicable publications, must be modified to fit the unique requirements of the HN forces being trained. Procedures may vary, but the fundamental techniques and thought processes still apply. Training assistance should focus on the materiel, fiscal, and logistical realities of the HN.

d. In general, those skills, concepts, and procedures for FID taught to US forces are also applicable to HN forces for IDAD. Training emphasis varies according to the HN requirements, force composition, and US and HN agreements. The training to be conducted depends on the situation and varies considerably. Existing military personnel, new military personnel, or paramilitary forces may receive training assistance.

e. HN counterpart personnel must be present with US trainers. The goal is for these counterparts to eventually conduct all instruction and training without guidance from US personnel. Initially, US personnel may present all or most of the instruction with as much HN assistance as is feasible. US trainers use the train the trainer concept.

f. Training assistance consists of all formal training conducted by joint force units. However, all personnel engaged in training assistance must be cognizant that they are typically under the sharp and sometimes magnified scrutiny of HN government personnel, military, media, and ordinary citizens. Part of preparing personnel for providing training assistance is making them aware of the less tangible elements of training assistance that can have a deep impact. Their words and actions at all times serve as examples of professionalism. Joint force personnel should know that in many HNs their mere presence alongside their counterparts often bolsters that counterpart's prestige within their organization and among the populace. Those providing training assistance should be aware that many HNs have a domestic PSYOP program and exploiting the presence of highly skilled US trainers may be part of their agenda. Generally, none of the less tangible offshoots of providing training assistance will be detrimental toward the joint force mission or US national policy as long as trainers are prepared for them.

g. **Advisory Assistance.** Within DOD, the principal element charged with providing advisory assistance is the SCO. The SCO includes all DOD elements, regardless of actual title, assigned in foreign countries to manage SA programs administered by DOD. The US advisor may often work and coordinate with civilians of other country team agencies. When he does, he must know their functions, responsibilities, and capabilities, because many activities cross jurisdictional borders. Together, the advisor and his counterpart must resolve problems by means appropriate to the HN without violating US laws and policies in the process. Advisors operate under very specific ROE to ensure that advisors remain advisors.

h. The joint force advisor must understand the scope of SCO activities. He also must know the functions, responsibilities, and capabilities of other US agencies in the HN. Because many joint force (notably SOF) activities cross the jurisdictional boundaries or responsibilities of other country team members, the advisor seeks other country team members to coordinate his portion of the overall FID effort.

i. In some situations, the HN may refuse US advisors. HN military leaders may instead request and receive other types of assistance such as air or fire support. To coordinate this support and ensure its proper use, US liaison teams accompany HN ground maneuver units receiving direct US combat support. Language-qualified and area-oriented SOF teams are especially suited for this mission. The HN government may refuse lethal support but eagerly accept support from engineer, medical, PSYOP, CA, and various intelligence capabilities.

j. **Trainer/Advisor Checklists.** The predeployment site survey (PDSS) leader— along with any subordinates he may specify—establishes effective initial rapport with the HN unit commander. The PDSS leader:

(1) Conducts introductions in a businesslike, congenial manner using the HN language.

(2) Briefs the HN commander on the joint force advisors' PDSS mission and the restrictions and limitations imposed on the unit by the higher US commander. The PDSS leader should use the HN language and, if required, visual aids translated into the HN language.

(3) Assures the HN commander that all PDSS team members are fully supportive of the HN's position and that they firmly believe a joint and HN-unit effort will be successful.

(4) Assures the HN commander that his assistance is needed to develop the tentative objectives for advisory assistance to include advisory team agreements with the HN commander on training objectives.

(5) Deduces or solicits the HN commander's actual estimate of his unit's capabilities and perceived advisory assistance and material requirements.

(6) Explains the PDSS team's initial plan for establishing counterpart relationships, obtains approval from the HN commander for the plan, and requests to conduct the counterpart linkup under the mutual supervision of the PDSS leader and the HN commander.

(7) Supervises the linkup between PDSS team members and their HN counterparts to determine if the HN personnel understand the purpose of the counterpart relationship and their responsibilities within it.

(8) Identifies reachback requirements.

(9) The PDSS leader should not make any promises or statements that could be construed as promises to the HN commander regarding commitments to provide the advisory assistance or fulfill material requirements.

k. The PDSS team members analyze the HN unit's status according to their area of expertise for the purpose of determining the HN requirements for advisory assistance. The PDSS team members:

(1) Explain the purpose of the analysis to counterparts.

(2) Encourage counterparts to assist in the analysis, the preparation of estimates, and the briefing of the analysis to the advisory team and HN unit commanders.

(3) Collect sufficient information to confirm the validity of current intelligence and tentative advisory assistance COAs selected prior to deployment.

(4) Collect and analyze all information relating to FP.

(5) Prepare written, prioritized estimates for advisory assistance COAs.

(6) Brief, with their counterparts, the estimates to the PDSS team and HN unit commander.

(7) Inspect, with their counterparts, the HN facilities that will be used during the assistance mission.

(8) Identify deficiencies in the facilities that will prevent execution of the tentatively selected advisory assistance COAs.

(9) Prepare written or verbal estimates of COAs that will correct the deficiencies or negate their effects on the tentatively selected advisory assistance COAs.

(10) Supervise the preparation of the facilities and inform the JFC of the status of the preparations compared to the plans for them.

l. Once received, the PDSS leader supervises the processing of the survey results. The PDSS leader then:

(1) Recommends to the HN unit commander the most desirable COAs emphasizing how they satisfy actual conditions and will achieve the desired advisory assistance objectives.

(2) Ensures that his counterpart understands that the desired COAs are still tentative contingent on the tasking US commander's decision.

(3) Selects the COAs to be recommended to the follow-on joint units, after obtaining input from the HN unit commander.

(4) Ensures the higher in-country US commander is informed of significant findings in the team survey for HN assistance.

m. The PDSS team plans its security in accordance with the anticipated threat. Adjustments are made as required by the situation on the ground. The PDSS team members:

(1) Fortify their positions (quarters, communications, medical, command) in accordance with the available means and requirements to maintain low visibility.

(2) Maintain a team internal guard system, aware of the locations of all other joint force advisors, and ready to react to an emergency by following the alert plan and starting defensive actions.

(3) Maintain a team internal alert plan that will notify all team members of an emergency.

(4) Maintain communications with all subordinate team members deployed outside of the immediate area controlled by the team.

(5) Establish plans for immediate team defensive actions in the event of an insurgent or terrorist attack or a loss of HN rapport with hostile reaction.

(6) Discuss visible team security measures with HN counterparts to ensure their understanding and to maintain effective rapport.

(7) Encourage the HN unit, through counterparts, to adopt additional security measures that have been identified as necessary during the analysis of the HN unit status and the inspection of its facilities.

(8) Establish mutual plans with the HN unit, through counterparts, for defensive actions in the event of an insurgent or terrorist attack.

(9) Rehearse team alert and defensive plans.

(10) Encourage the HN unit, through counterparts, to conduct mutual, full-force rehearsals of defensive plans.

n. **Executing the Mission.** The senior joint force advisor assists the HN unit commander in providing C2 during the execution of the operation. Accompanying an HN commander on missions will afford the advisor visibility on the interactions between the HN forces and the populace. The senior advisor:

(1) Monitors the tactical situation and recommends changes to the present COA to gainfully exploit changes in the situation.

(2) Monitors the location of the HN commander and recommends changes so that he can provide leadership at critical points and not deprive himself of the ability to maneuver his force in response to tactical changes.

(3) Monitors the information flow to the HN commander and recommends improvements needed to:

(a) Make continuous use of intelligence-collection assets.

(b) Keep subordinates reporting combat information.

(c) Screen the information given to the HN commander to prevent information overload.

(d) Keep the command communications channels open for critical information.

(4) Monitors the HN commander's control of the execution and recommends improvements to:

(a) Focus combat power on the objective.

(b) Keep movement supported by direct and indirect fire.

(c) Maintain mutual support between subordinate elements.

(d) Maintain fire control and discipline.

(e) Consolidate and reorganize during lulls in the battle or after seizing the objective.

(f) Conduct effective stability operations to gain popular support.

(g) Support progress in such areas as governance, infrastructure, humanitarian assistance, and economic development.

(h) Support activities with effective IO.

(5) Monitors any command succession and assists the new HN unit commander to smoothly and rapidly take control of the execution of the operation.

o. The joint force advisory team members also assist their counterparts during the execution of the operation. The advisory team members:

(1) Monitor staff functions and recommend improvements or corrections, as needed.

(2) Monitor the technical or tactical execution of individual tasks and recommend improvements or corrections, as needed.

(3) Remain continuously aware of the tactical situation.

(4) Execute joint force unilateral contingency plans, as required by the situation.

Note. Document recurring or significant problems or events for reference during end-of-mission debriefings and reports.

p. The advisory team presents the instruction. Trainers/advisors:

(1) Adhere to the lesson outlines consistent with the cooperation from the HN forces and changes in the mission, enemy, terrain and weather, troops and support available, time available, and civil considerations.

(2) State clearly the task, conditions, and standards to be achieved during each lesson at the beginning of the training (to include training exercises) and ensure the HN students understand them. (Human rights should be emphasized in the appropriate period of instruction.)

(3) Demonstrate the execution or show the desired end result to clearly illustrate the task.

(4) Stress the execution of the task as a step-by-step process, when possible.

(5) Monitor the HN students' progress during practice and correct mistakes as they are observed.

(6) State (at a minimum) all applicable warning and safety instructions in the HN language.

(7) Monitor periodically instructions given through HN interpreters to ensure accurate translations using HN-language-qualified joint force personnel.

q. The joint force ensures the security of the training sites. Advisors or designated security personnel:

(1) Analyze the threat to determine any capabilities to attack or collect intelligence on the HN unit's training at each site.

(2) Prepare estimates of COAs that would deny the training sites to the insurgents or terrorists.

(3) Recommend to the HN unit commander that he order the adoption of the most desirable COA, stressing how it best satisfies the identified need.

(4) Ensure before each training session (using, as a minimum, brief back rehearsal) that all personnel—both US and HN—understand the defensive actions to be taken in the event of an insurgent or terrorist attack and any OPSEC measures to be executed.

r. Designated advisory/training team members maintain written administrative training records. These members:

(1) Encourage HN counterparts to assist.

(2) Record all HN personnel and units who receive training and identify the type of training they receive.

(3) Organize records to identify training deficiencies and overall level of HN proficiency.

(4) Identify specific HN personnel or units who demonstrate noteworthy (good or bad) performance.

(5) Identify to the joint force and HN unit commanders the noted training deficiencies, noteworthy performances, and required additional or remedial training.

Intentionally Blank

CHAPTER VI
FOREIGN INTERNAL DEFENSE OPERATIONS

"The war on terrorism will be fought with increased support for democracy programs, judicial reform, conflict resolution, poverty alleviation, economic reform, and health and education. All of these together deny the reason for terrorists to exist or to find safe haven within borders."

Colin Powell
Secretary of State
United Nations Security Council, 12 November 2001

SECTION A. EMPLOYMENT CONSIDERATIONS

1. General

a. Thus far, the discussion of FID has generally centered on the strategic and operational levels. This chapter transitions to a more focused examination of the employment principles, tools, and techniques used in conducting FID operations.

b. **FID activities are part of the unified actions of the combatant command and emphasize interagency coordination.** Even small tactical operations will usually require interagency coordination, most likely through the SCO.

2. Employment Factors

As in planning, several areas deserve special attention when discussing employment of forces in FID operations (see Figure VI-1).

a. **Information Operations Impact.** IO involve actions taken to affect adversary information and information systems while protecting one's own information and information systems. IO apply across all phases of an operation, throughout the range of military operations. IO and related activities affect the perceptions and attitudes of adversaries and a host of others in the operational area. During FID operations, IO disciplines must be closely integrated in all aspects of planning and execution.

b. **Psychological Impact.** Regardless of where or when PSYOP is conducted, the FID and IDAD objectives are kept in mind. The impact of these efforts may occur incidentally, as a result of another unrelated operation, or may be the result of an operation specifically executed for its psychological effect.

c. **Intelligence Support.** A thorough intelligence analysis must focus on the political, social, scientific, technical, medical, and economic aspects of the area as well as on an analysis of hostile elements. Active intelligence support must continue through to the end of the employment of military forces. This continuous intelligence effort will gauge the reaction of the local populace and determine the effects on the infrastructure of

FORCE EMPLOYMENT FACTORS IN FOREIGN INTERNAL DEFENSE OPERATIONS

INFORMATION OPERATIONS IMPACT

Information operations disciplines must be closely integrated in all aspects of foreign internal defense (FID) planning and execution.

PSYCHOLOGICAL IMPACT

The psychological effort is relevant to the entire FID operation.

INTELLIGENCE SUPPORT

A thorough intelligence analysis must focus on the political, social, scientific, technical, medical, and economic aspects of the area as well as on an analysis of hostile elements.

FORCE SELECTION

Success is most effectively achieved through employing operational designs that provide a combination of conventional forces while leveraging the unique capabilities of special operations forces.

PUBLIC INFORMATION PROGRAMS

Public information is an important ongoing effort during the employment phase of any FID mission.

LOGISTIC SUPPORT

Logistic operations in support of FID are both supporting missions to United States forces and primary operational missions when supporting host nation civilians or military forces with medical, construction, maintenance, supply, or transportation capabilities.

COUNTERDRUG OPERATIONS IN FOREIGN INTERNAL DEFENSE

United States military support of the national counterdrug effort has increased tremendously in recent years.

COUNTERTERRORISM AND FOREIGN INTERNAL DEFENSE

Subversion, lawlessness, and insurgency can all contribute to the growth of terrorists and terrorism.

OPERATIONS SECURITY

A major problem in all FID activities is denial of critical information about friendly intentions, capabilities, and activities to hostile elements.

LESSONS LEARNED

As FID activities are conducted, it is critical to document lessons learned to allow the commander to modify future operations and activities to fit the special circumstances and environment.

Figure VI-1. Force Employment Factors in Foreign Internal Defense Operations

US efforts as well as evaluate strengths, weaknesses, and disposition of opposition groups in the area.

(1) Appendix B, "Joint Intelligence Preparation of the Operational Environment to Support Foreign Internal Defense," provides intelligence considerations and a format for intelligence preparation of the FID operational area. Although the considerations must be modified for the specific FID operation, generally the operational area must be

surveyed for an OAE; a geographic, population, and climatology analysis; and a threat evaluation. These factors will dictate the employment techniques and FID tools to use.

For further information on JIPOE, refer to JP 2-01.3, Joint Intelligence Preparation of the Operational Environment.

(2) **Human intelligence (HUMINT) resources support is likely to be the most important type of intelligence support of FID.** The analysis requires a large HUMINT effort. This requires a more decentralized intelligence gathering approach than in conventional operations. Small units and teams deployed in the operational area are in a good position to evaluate the social, economic, and military situation in the HN. The best US intelligence may come from units and teams that work closely with the local population and HN military forces. These units and teams must be prebriefed and debriefed for priority intelligence requirements (PIRs), and this information should contribute to the data used in conducting the intelligence preparation of the operational environment.

(3) **Information sharing across USG and national boundaries is an important concept in FID.** There are likely to be several government agencies operating in an HN, and all are exposed daily to information valuable to FID operations and the success of the IDAD program. This requires a strong focus on the development of an effective process for interagency information exchange and coordination. In addition, the very nature of FID denotes the sharing of information between the supported HN and the US joint force headquarters controlling the FID effort. This information exchange may be further complicated by a friendly third nation participating in FID operations or the HN's IDAD program.

For further information on interagency coordination, refer to JP 3-08, Interorganizational Coordination During Joint Operations.

(4) **The nature of FID missions and the high degree of dependence on HUMINT sources necessitate an active CI and OPSEC program.** At a minimum, US forces must be able to:

(a) Accomplish liaison with HN CI and security forces. Preestablished relationships with the country team are necessary and should be regularly maintained in advance of FID operations, wherever practical, as they will maximize efficiency and effectiveness.

(b) Provide a conduit to country team CI and security elements.

(c) Conduct analysis of opposing force intelligence collection, security, CI, and deception capabilities, and propaganda.

(d) Conduct the full range of CI operations.

(e) Conduct CI vulnerability assessments of US forces.

(f) Provide CI input to US FID plans.

(5) US force deployments for FID missions must be structured to provide adequate CI resources and plan for reachback to the national strategic CI community to accomplish these missions.

d. **Force Selection.** US forces in general have some ability to assess, train, advise, and assist foreign forces. The degree to which they can be tasked to do so depends on their preparation in terms of language and other skills and the knowledge necessary to function within the operational environment. **Success is most effectively achieved through employing operational designs that provide a combination of CF, while leveraging the unique capabilities of SOF.** The selection of the appropriate ratio of SOF and CF forces must be a deliberate decision based on thorough mission analysis and a pairing of available capabilities to requirements. The most important factor informing this decision is the capability and expertise required rather than the size of the force required. Additional factors include the political sensitivity of the mission and requirements for cultural and language experts or other special requirements. JFCs must be aware that operations may change rapidly in character, and that their force structures may need to adapt as well. Both the integration of SOF with CF and vice versa are increasingly the norm.

(1) **Special Operations Forces.** SOF may conduct FID operations unilaterally in the absence of any other military effort, support other ongoing military or civilian assistance efforts, or support the employment of CF.

(a) SOF units possess specialized capabilities for FID, including support for COIN and, when applicable, for unconventional warfare. Other support includes CAO, PSYOP support, and training in specific areas, typically with HN SOF. In addition, SOF may support combat operations by conducting highly specialized missions. However, the typical SOF role in FID is to train, advise, and support HN military and paramilitary forces.

(b) In addition to the specific capability requirements that may call for selection of SOF, the nature of the FID mission itself may dictate the use of SOF. SOF's unique capabilities for language, cultural awareness, regional focus, etc. may be required when the environment involves particular political sensitivities. Additionally, SOF's ability to conduct short-notice missions, with only modest support, makes them adept at initiating programs for hand-over to CF.

(c) USSOCOM provides SOF in support of GCCs. SOF contribute to the FID effort normally under OPCON of the theater SOC, which has primary responsibility to plan and supervise the execution of SOF operations in support of FID. SOF also provide dedicated theater forces. When planning for use of SOF, command, control, communications, and computers requirements among the combatant command, the country team, and SOF must be assessed. Communications requirements for C2, administration, logistics, and emergencies must be clarified.

(2) **Conventional Forces.** When the FID effort requires broader action to support HN IDAD efforts, the JFC may predominantly employ CF in the FID mission. CF contain and employ organic capabilities to conduct FID indirect support, direct support, and combat operations. This may include providing intelligence and logistic support to HN units, serving as military advisors, conducting MTTs, embedding US units into HN units, conducting joint/multinational operations with HN units, and serving as a quick reaction force in support of HN units. Unit commanders must be given clear guidance on unit mission requirements that include the need to prepare their forces to conduct FID. USCG training teams, personnel, and platforms are well suited to support the development of stable, multi-mission maritime regimes to respond to many transnational threats. USCG FID activities reach beyond normal military-to-military relations to a broader HN maritime audience, including, but not limited to, law enforcement agencies, maritime administrations, and transport ministries.

e. **Public Information Programs.** Public information is important during all phases of any FID mission. While it is important to correctly portray the FID effort to HN personnel through PSYOP, it is also important to employ an effective PA program to inform HN and US publics of current FID actions, goals, and objectives. History has shown that without popular support, it may be impossible to develop an effective FID operation. At the US national level, public diplomacy programs will accurately depict US efforts. This national program is supported through the CCDR's (or subordinate JFC's) information programs designed to disclose the maximum amount of information possible within applicable security restrictions and the guidelines established by the President or SecDef. Coordination is essential between the PA staff and the media, the country team, the PSYOP element, and other information agencies within the HN and region.

f. **Logistic Support.** Logistic operations in support of FID **support both US forces and primary operational missions** (supporting HN civilians or military forces with medical, construction, maintenance, supply, or transportation capabilities). General guidelines for logistic issues in support of US forces conducting FID operations include:

(1) There may be a ceiling imposed on the number of US military personnel authorized to be in the HN to conduct FID operations. Commanders should determine how sea basing forces impacts this decision. Maximum use should be made of host-nation support (HNS) capabilities, but where reliance on the HN is not feasible, logistic support requirements must be minimized. FID and its support may include contractor personnel, which could complicate legal, diplomatic, administrative, budgetary, and logistical issues. Efficient use of throughput of supplies (an average quantity that can pass through a port on a daily basis), airlift resupply, and inter-Service support agreements should also be considered.

(2) Commanders must carefully balance the advantages of using HNS with the danger of establishing dependence on potentially unreliable sources.

(3) Logistic operations are tailored to the type of mission. Service logistic support elements will be integrated into the overall joint force. Logistic support for the deployed forces, however, will remain a Service responsibility.

(4) HNs often require support beyond their organic capabilities. Accordingly, **when conducting FID with multinational partners, there becomes a need to establish multinational logistic support agreements.** The need for such non-organic support must be identified during the planning phase of FID support and arranged for prior to participation in the operation. Acquisition and cross-servicing agreements (ACSAs) negotiated with multinational partners are beneficial to the FID effort in that they allow US forces to exchange most common types of support. Authority to negotiate these agreements is usually delegated to the GCC by SecDef. Authority to execute these agreements lies with SecDef and may or may not be delegated.

For further information on international logistics, refer to JP 4-08, Multinational Logistics.

g. **Counterdrug Operations in FID.** Narcotics production and trafficking can flourish in countries where subversion, lawlessness, and insurgency exist. Accordingly, **FID operations complement CD efforts by reducing those problems in partner nations.** CD focused programs are integrated into theater strategies as a coordinated effort to support HN governments' IDAD strategies.

(1) DOD resources may be used in connection with CD activities in nations receiving military assistance in support of an IDAD program. This military assistance is often centered on source operations, but can be involved with in-transit CD operations.

(a) **DOD is the lead agency of the USG for the detection and monitoring (D&M) of aerial and maritime transit of illegal drugs into the US.** This mission is performed with O&M funds, notwithstanding the possibility of incidental benefit to the HN. Such activities may include nonconfrontational intercepts for intelligence or communication purposes and gathering and processing of tactical intelligence from a variety of sources, including fixed and mobile surveillance assets and certain intelligence sharing.

(b) In a CD support role (subject to national policy and legislative guidance) DOD may offer certain direct support to HN CD personnel, and certain enhanced support to US civilian law enforcement agencies that may be operating in the HN, and to DOS, INL.

(2) Absent direction from SecDef, DOD forces engaged in CD activities are prohibited from engaging in direct law enforcement activity. They may not directly participate in an arrest, search, seizure, or other similar activity. DOD personnel are not

authorized to accompany HN forces on actual CD field operations or participate in any activities where hostilities are likely to occur. USN ships contribute significantly to the D&M phase of CD operations, as they are frequently in a position to intercept and apprehend maritime drug smugglers. Because DOD does not directly participate in search, seizure, arrest, and other similar activities, USCG law enforcement detachments, who are authorized to perform law enforcement activities, are frequently embarked in USN and allied ships to act in this capacity as prescribed in Title 10, USC, Section 379.

(3) As directed by SecDef through the CJCS, GCCs will be given the authority to plan and execute HN programs using a combination of SA, training and advisory assistance (non-SA-funded), intelligence and communications sharing, logistic support, and FHA. These efforts are designed to bolster the HN's capability to operate against the infrastructure of the drug-producing criminal enterprises.

(4) CCDRs and subordinate JFCs must coordinate closely with the country team DEA and DOS international narcotics matters representatives. Liaison with the Office of National Drug Control Policy (ONDCP) is also vital. ONDCP is legislatively charged with the responsibility of establishing the national drug control strategy and with coordinating and overseeing the implementation of the consolidated National Drug Control Program budget. This coordination is crucial to an efficient national CD program to combat illicit drug trafficking in source regions.

For further information on joint CD operations, refer to JP 3-07.4, Joint Counterdrug Operations.

h. **Combating WMD Operations in FID.** Combating weapons of mass destruction (CWMD) is one of the greatest challenges facing the US. FID operations have significant potential to build our partners' capacity to combat WMD.

For further information on CWMD operations, refer to JP 3-40, Combating Weapons of Mass Destruction.

i. **Counterterrorism and FID.** Subversion, lawlessness, and insurgency can all contribute to the growth of terrorists and terrorism. FID can complement CT by reducing these contributing factors. **Specific AT and CT efforts can be conducted during FID operations in support of an HN's IDAD program.**

(1) Enhancing the will of other states to fight global terrorism primarily is the responsibility of DOS. Effective FID, however, can improve public perceptions of the HN and USG and facilitate more active HN policies to combat terrorism. More directly, military-to-military contacts can help make HN officials advocates of potential operations against terrorist capabilities.

(2) In many cases, measures increasing the capacity of a state to fight terrorism also will strengthen its overall IDAD program. These measures can include the following:

(a) Developing the ability of the HN financial transactions, break funding streams for criminal and insurgent groups, and prosecute their members. This may involve greater US-HN cooperation in developing regulated financial institutions.

(b) Ensuring that HN security personnel have access to appropriate equipment and training to conduct all phases of combating terrorism operations.

(c) Training personnel at entry and exit points (including airports, seaports, and border crossings) to identify and apprehend individuals and materials being used by international terrorist groups.

(d) Assisting HN security and intelligence agencies to be included into international networks that can share information on terrorist activities.

(e) Developing effective judicial systems, and minimizing corruption and intimidation of HN officials.

For further information on COIN, refer to JP 3-24, Counterinsurgency.

For further information on CT, refer to JP 3-26, Counterterrorism.

j. **Operations Security.** A major problem in all FID activities is denial of critical information about friendly intentions, capabilities, and activities to hostile elements. The nature of FID implies that many HN officials and populace will know of certain US activities as they occur. Criminal and insurgent groups may have members or sympathizers within HN institutions that could be informants. US and foreign personnel involved in FID and IDAD programs should be provided extensive OPSEC training to ensure effectiveness of their operations.

For further information on OPSEC, refer to JP 3-13.3, Operations Security.

k. **Lessons Learned.** As FID operations are conducted, it is critical to document lessons learned to allow the commander to modify future operations and activities to fit the special circumstances and environment. Comprehensive after-action reviews and reports focusing on the specifics of the FID operations should be conducted to gather this information as soon as possible after mission execution. The Joint Center for Operational Analysis within the United States Joint Forces Command (USJFCOM), the Services, and other government agencies' (OGAs') lessons learned programs provide readily available sources of information to FID planners and operators. In addition, USSOCOM's Special Operations Debrief and Reporting System, an internal USSOCOM-only program, also can provide additional information on peacetime FID missions.

For further information for specific reporting procedures, refer to CJCSI 3150.25D, Joint Lessons Learned Program.

3. Site Survey Considerations

Units assigned a FID mission must implement procedures to help DOS and the country team vet HN forces before they can receive training. Any personnel who are not vetted must be removed from training. The primary purpose of vetting is to ensure the identification of personnel with a history of human rights violations. US policy as well as the 1978 "Kennedy Amendment" to the FAA (Title 22, USC, Section 2304[a][2]) prevents US cooperation with and SA funding to a government of any country that engages in a consistent pattern of gross violations of internationally recognized human rights. Ideally, a site survey team gathers this information. Such teams should include a CI representative, preferably an FPD or foreign area officer (FAO) asset. To properly conduct the training, units assigned to FID operations need to determine or identify:

a. The HN unit mission and its mission-essential task list and its capability to execute them.

b. The organizational tables for authorized personnel and equipment and for personnel and equipment actually on hand.

c. Any past or present foreign military presence or influence in doctrine, training, or combat operations.

d. The unit's ability to retain and support acquired skills or training from past MTTs or foreign training missions.

e. The organization and leadership level that is responsible for training the individual soldier. Does the HN have institutional training established? Is it effective?

f. Any operational deficiencies during recent combat operations or participation in joint or multinational exercises with US personnel.

g. The maintenance status, to include maintenance training programs.

h. The language or languages in which instruction will be conducted.

i. The religious, tribal, or other affiliations within the HN forces that need to be considered (notably the differences between HN forces and the local populace).

j. The potential security concerns with employing US members (and allies) in the HN training areas.

k. The local infrastructure and possible positive or negative impacts of training on the local populace.

l. The local populace's attitudes toward US military and government personnel, as well as ordinary US citizens (to include presence and behavior of expatriate US populations).

m. The local populace's prejudices or fears.

n. Any key local leaders, communicators, and potential spoilers.

o. The presence, agendas, capabilities, influence, and attitudes of NGOs and IGOs.

p. The FID operations unit's area assessment needs. For example - local media for deploying PSYOP.

4. Health Service Support and Medical Civil-Military Operations

a. US joint medical personnel and forces can be employed as indirect support or direct support during a FID operation. The predominant types of engagement applied will depend on the make up of the HN military medical forces and the HN civilian health sector and their respective roles in that nation's IDAD program. For US joint medical forces, health engagement will include varying degrees of military-military activities and medical civil-military operations (MCMO). In some countries the military and civilian health systems may be completely separate while in other nations the two systems may be integrated, necessitating a unified approach. Medical input and involvement in indirect support to FID such as SA, exchange programs, and multinational exercises must be provided at the onset of planning and targeted toward the health problems facing the HN military and, in conjunction with other US agencies, civilian health initiatives through CAO and FHA. Possible MCMO activities during FID operations include:

(1) Providing public health activities, to include preventive medicine and veterinary care, food hygiene, immunizations of humans and animals, childcare, preventive dental hygiene, and paramedic procedures. This includes:

(a) Medical civic action programs.

(b) Dental civic action programs.

(c) Veterinary civic action programs.

(2) Providing triage, diagnostic, and treatment training.

(3) Developing logistic programs.

(4) Developing education programs.

(5) Developing HSS intelligence and threat analysis.

(6) Developing HN military field HSS support system for treatment and evacuation.

(7) Assisting in the upgrade, staffing, and supplying of existing HSS facilities.

(8) Developing wellness and preventive care, including public information programs.

b. Military forces should be employed in MCMO missions that are affordable and sustainable by the HN. This includes pursuing realistic training and acquisition programs. In addition to training HN personnel during FID operations, medical education opportunities for HN personnel through IMET may be pursued. Following a course of realistic HSS measures and programs may also entail mitigating unrealistic expectations among the HN populace. Other second order effects can emerge from HSS as well, such as a real or perceived imbalance in health care development. Resources should be shifted to areas where imbalance exists.

c. FID units typically can provide only a small portion of the HN's HSS needs; therefore, close cooperation with OGAs, IGOs, and NGOs can enhance the support provided by the military. Commanders should seek to increase the effectiveness of other USG agency programs such as USAID whenever possible. Working with or near IGOs and NGOs may be untenable due to their desire to preserve the perception of neutrality. Military units may have to settle for awareness of IGO and NGO activities and employ themselves so as not to duplicate efforts in HSS.

For further information on HSS, refer to JP 4-02, Health Service Support.

For information on MCMO, refer to JP 3-57, Civil-Military Operations

SECTION B. INDIRECT SUPPORT

5. General

a. This category of support provides equipment or training support in order to enhance the HN's ability to conduct its own operations.

b. Three subsets of indirect support include SA, joint and multinational exercises, and exchange programs. (See Figure VI-2.)

6. Security Assistance

a. This section will discuss specific military SA operations and how the GCC may use this tool to further support FID activities. **The military will primarily provide equipment, training, and services to the supported HN forces.** In the SA arena, GCCs and subordinate JFCs do not have authority over the SA program, but have responsibility for planning and executing military activities to support FID within the SA process.

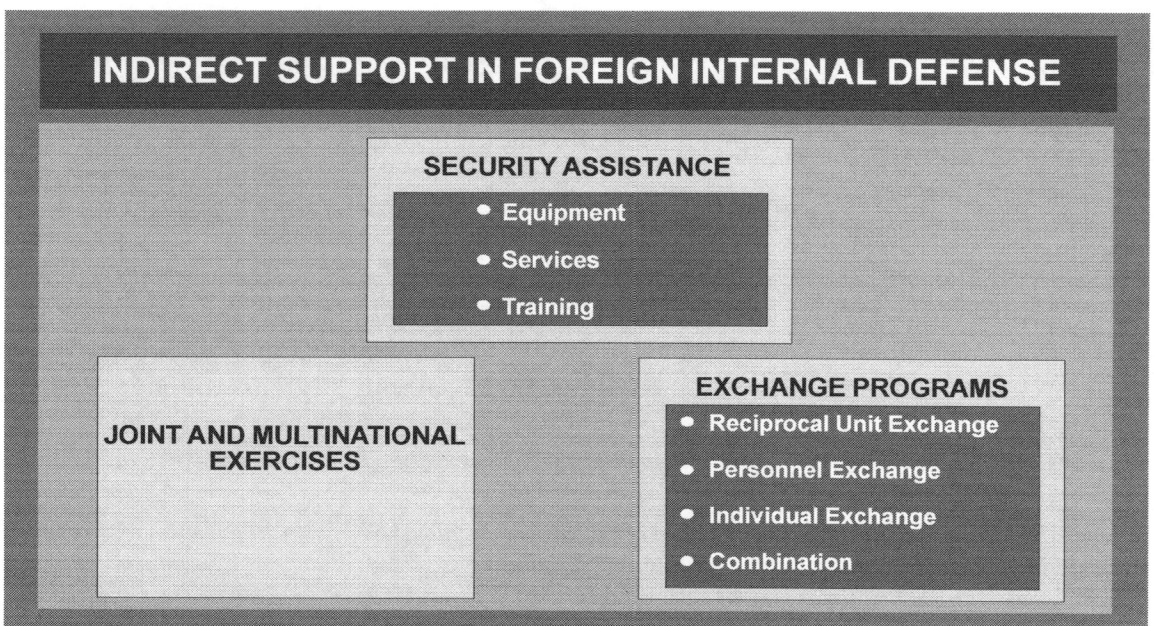

Figure VI-2. Indirect Support in Foreign Internal Defense

GCCs are active in the SA process by advising ambassadors through the SCO and by coordinating and monitoring ongoing SA efforts in their AOR. In addition, through coordination with HN military forces and supporting SCOs, the GCC can assist in building credible military assistance packages that best support long-term goals and objectives of regional FID operations. The following paragraphs describe the SA support areas of equipment, services, and training as well as the employment considerations for each.

b. **Equipment.** The GCC and subordinate JFC can have the greatest impact in this area during the planning and resource identification phase of developing the theater strategy. **Regional threats identified and level of HN technology will determine the general equipment needs of the supported HNs in the theater.** Each SCO will coordinate resultant military equipment requests with the GCC's staff and country team. Finally, the HN provides a letter of request to the SCO, who forwards it to the appropriate Military Department to determine price and availability. Throughout this process, HN needs must be evaluated in terms of the threat and existing social, political, and economic conditions. Care must be taken to guard against a US solution or to support unnecessary requests, as explained below.

(1) **The FID planning imperative to tailor support to HN needs is extremely important in providing equipment support.** Environmental factors, level of HN training, ability to maintain equipment, HN infrastructure, and a myriad of other factors will determine what equipment is appropriate to the HN's needs. If equipment in the US inventory is not appropriate for use by the HN, the commander may recommend a nonstandard item to fill the requirement. Sustainability of nonstandard equipment, as well as interoperability with existing equipment, must be considered.

(2) HNs may request expensive equipment as a status symbol of regional military power. This is often done in spite of the fact that the overall strength of the military would be best enhanced by improved training and professionalism among the existing force. This is a delicate political situation, but one that the ambassador and the GCC may be able to influence.

c. **Services.** Services include any service, test, inspection, repair, training, publication, technical or other assistance, or defense information used for the purpose of furnishing military assistance, but does not include military education and training activities. **Services support is usually integrated with equipment support.** The CCDR has oversight to ensure that the equipment is suitable for HN needs and that the HN is capable of maintaining it. These types of services will almost always be required to ensure an effective logistic plan for the acquired equipment. There are two common types of service teams: QATs and TATs. QATs are short term and are used to ensure that equipment is in usable condition. TATs are used when the HN experiences difficulty with US-supplied equipment. For detailed information on teams available for initial and follow-on equipment support, see DOD Manual 5105.38-M, *Security Assistance Management Manual*.

d. **Training.** The training portion of SA can make a very significant impact on the HN IDAD program. **The GCC is actively involved in coordinating, planning, and approving training support with the SCO and HN.** The Services, through their SA training organizations, are the coordinators for SA-funded training.

(1) The following are the general objectives of training programs under SA (see Figure VI-3).

(a) **Professional Military Education.** To further the goal of regional stability through effective, mutually beneficial military-to-military relations which culminate in increased understanding and defense cooperation between the United States and the HN.

(b) **O&M Skills.** To create skills needed for effective O&M of equipment acquired from the US.

Equipment that is delivered to a host nation must address needs in terms of the threat as well as the existing social, political, and economic conditions.

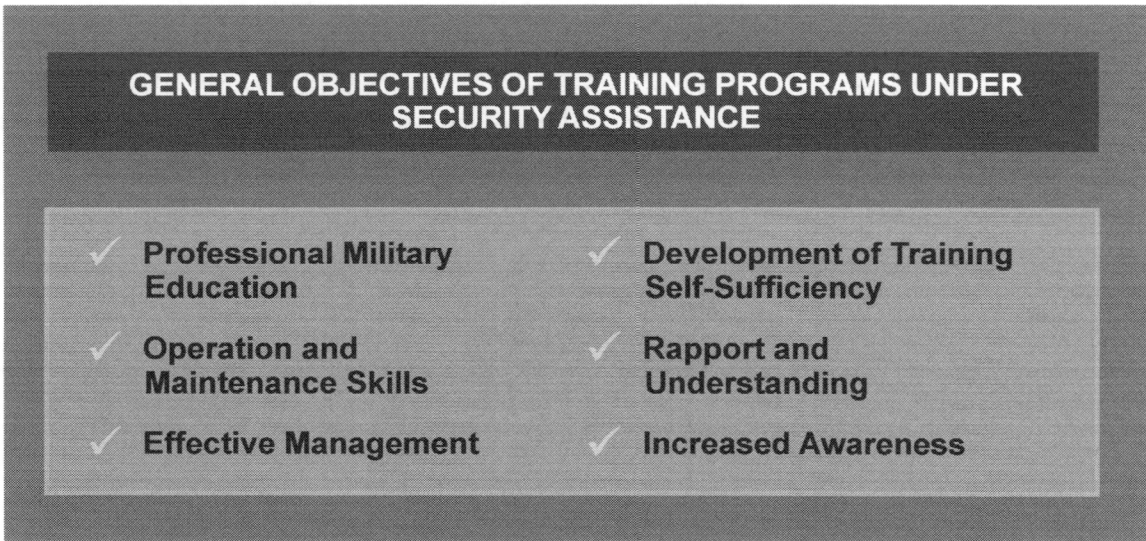

Figure VI-3. General Objectives of Training Programs Under Security Assistance

(c) **Effective Management.** To assist the foreign country in developing expertise and systems needed for effective management of its defense establishment.

(d) **Development of Training Self-Sufficiency.** To foster development by the HN of its own training capability.

(e) **Rapport and Understanding.** To promote military-to-military understanding leading to increased standardization and interoperability.

(f) **Increased Awareness.** To provide an opportunity to demonstrate the US commitment to the basic principles of internationally recognized human rights.

(2) The following force structure, training plan, and training activities considerations should be understood before implementing an SA training program.

(a) **Training Force Structure.** Services administer SA training with a combination of military and contracted trainers. SOF may be preferred for some types of training if the situation requires language or regional skills. HN forces can also receive education or training in US schools or contractor facilities.

(b) **Training Plan.** The training portion of SA is identified and coordinated by the SCO with the HN military. GCCs and subordinate JFCs will incorporate the SA training, planning, and requirements into the overall military planning to support FID operations.

1. The SCO will develop a 2-year plan, known as the combined training and education plan (CTEP), which consolidates HN needs from a joint perspective, taking into consideration all sources of training and funding. These plans will be approved by the appropriate combatant command.

<u>2.</u> Each year the GCCs will host an SC education and training working group. During this working group the SCO and Service training coordinators, along with the GCC, DOD, and DOS will refine and coordinate the previously approved CTEP, finalize the budget year training program, and announce and discuss changes in the command's training policy or procedures.

<u>3.</u> The SCO must ensure that the HN looks beyond its current needs toward the future. A tool to accomplish this is the 2-year plan.

<u>4.</u> Commanders and their staffs must also ensure that US forces involved in providing SA-funded training to HN personnel in-country are fully aware of restrictions on their involvement in HN combat operations and that they employ vigilant FP measures. US training teams should be considered likely targets of attack if supported forces are facing an active armed threat.

(c) **Training Activities.** The GCC has a number of training activities that should be considered when reviewing individual country training plans. These programs may be carried out by HN personnel attending military schools in the US or by deploying teams of SOF, CF, or a combination of both. Consideration should be given to language capabilities, cultural orientation, theater objectives, complexity of tasks/missions to be performed, and the supported HN's IDAD program when selecting forces. The following are the primary types of training that may be employed as part of military support to the SA program.

<u>1.</u> **Mobile Training Teams.** MTTs are used when an HN element requires on-site training and to conduct surveys and assessments of training requirements. An MTT may be single-Service or joint, SOF or CF, but is tailored for the training the HN requires. An MTT is employed on a temporary duty (TDY) basis for a designated period of time. If HN forces require training for a longer period, training in the United States should be considered as an alternative.

<u>2.</u> **Extended Training Service Specialists (ETSSs).** ETSS teams are employed on a permanent change of station (PCS) basis (usually for 1 year) in order to assist the HN in attaining readiness on weapons or other equipment. These teams train the HN's initial instructor cadre so that they can assume the responsibility for training their own personnel. They consist of DOD military and civilian personnel technically qualified to provide advice, instruction, and training in the installation, operation, and maintenance of weapons, equipment, and systems to HN personnel.

<u>3.</u> **Technical Assistance Field Teams.** TAFTs consist of DOD personnel in a PCS status, assigned to provide technical or maintenance assistance to HN personnel. TAFTs are also deployed on a PCS basis and train HN personnel in equipment-specific military skills.

<u>4.</u> **International Education and Training.** The US international education and training program provides HN personnel with military education and

training opportunities in the US. This type of training not only meets the immediate HN requirement of increased training, but also has a longer term impact of improving US-HN relations.

5. **Contractors.** Contractor personnel can be utilized in the execution of SA as outlined in the DOD 5105.38-M, *Security Assistance Management Manual*. The provisions of DOD Instruction (DODI) 3020.37, *Continuation of Essential DOD Contractor Services During Crises*, should be fully understood should a situation escalate from indirect to direct support.

6. **The Regional Defense Combating Terrorism Fellowship Program (RDCTFP).** RDCTFP is a critical tool for DOD to provide grant aid education and training to support regional nations in our collective efforts in support of combating terrorism. RDCTFP offers education and training to foreign military officers, ministry of defense civilians, and other foreign civilian security force personnel. The DOD funded program is implemented through the US Services' SA training management systems.

(3) **Medical Civil-Military Operations in Indirect Support.** MCMO indirect support to FID operations is generally accomplished by medical training teams and advisors. The focus is on identification of health threats that affect the efficiency and effectiveness of the HN military forces and designing programs to train and equip those forces. Typically, the main effort of such training has been conducted by SF with support from other SOF units. PSYOP has historically supported and advised HN counterparts in programs and series to support MCMO missions. Wellness and instructional PSYOP series can seek to simply increase participation in medical or veterinary programs that some TAs may be reluctant to use due to cultural bias. In addition, introduction of new behaviors such as sanitary food or water practices can be pursued as well. This type of support can cross boundaries into the realm of direct support as well and may involve simultaneously conducting both indirect and direct support.

7. **Joint and Multinational Exercises**

These exercises can enhance a FID operation. **They offer the advantage of training US forces while simultaneously increasing interoperability with HN forces and offering limited HN training opportunities.** The participation of US forces in these exercises, primarily designed to enhance the training and readiness of US forces, is funded by O&M funds of the providing Service or USSOCOM if SOF are involved. Airlift and sealift may be provided by the combatant command from its airlift and sealift budget. Certain expenses of HN forces participation may be funded by the developing country multinational exercise program as arranged by the conducting combatant command. **These expenses differ from SA funding because SA is designed to train HN forces, whereas multinational and selected joint exercises are designed to train US forces in combination with HN forces.** Legal restrictions on what FID activities can be conducted in conjunction with these exercises are complex. Appendix A, "Legal Considerations," provides general guidelines on these restrictions. Prior legal guidance is important to the concept of the exercise and related FID operations. Exercises should be planned as part of the overall training program for the theater, and other FID activities should be integrated into the framework of these exercises. Examples of this integration are found in the conduct of HCA missions. The implementation of HCA programs into exercises will be examined in detail later in this chapter. Multinational and selected joint exercises can yield important benefits for US interests and the overall theater FID operation. The most significant of these benefits:

    a. Enhance relationships and interoperability with HN forces;

    b. Demonstrate resolve and commitment to the HN; and

    c. Familiarize US forces and commanders with HN employment procedures and potential combat areas.

8. Exchange Programs

    a. These programs allow the commander to use O&M money for the exchange of units or individuals and may be used to expand the efforts of the SA programs funded under IMET that allow HN personnel to train in the US. **These exchange programs foster greater mutual understanding and familiarize each force with the operations of the other.** Exchange programs are another building block that can help a commander round out the FID plan. These are not stand-alone programs; however, when commanders combine them with other FID tools, the result can be a comprehensive program that fully supports the HN IDAD program. The general types of exchange programs that commanders should consider are described below. Appendix A, "Legal Considerations," provides a more detailed explanation of the legal aspects of these types of training.

    b. **Reciprocal Unit Exchange Program.** This program is for squad-to-battalion-size elements. Each nation's forces trains the other in tactics, techniques, and procedures. This program is a good vehicle for US commanders to use in order to sensitize their forces to the cultural and social aspects of the HN while simultaneously increasing the training readiness of HN forces. The proficiency of the units must be comparable to preclude exchanging fully trained US forces for untrained HN forces.

    c. **Personnel Exchange Program (PEP).** The PEP is a 1- to 3-year program in which one person from the HN is exchanged with a US member. This program, like reciprocal unit exchanges, requires that the exchanged personnel be of comparable proficiency in their area of expertise.

    d. **Individual Exchange Program.** This program is similar to the PEP. It is different, however, because it is a TDY assignment in theater. This program gives the commander flexibility, since personnel will not be lost for extended periods and the commander is able to expose a larger portion of the force to the program.

e. **Combination Programs.** Commanders should consider combining SA efforts with joint or multinational exercises in order to obtain maximum benefit for all concerned. For example, exchange of key personnel during exercises will gain more in terms of interoperability than exchanges during normal operational periods. Also, the exchange of units with similar equipment, especially if the HN is unfamiliar with the equipment, may be very beneficial.

SECTION C. DIRECT SUPPORT (NOT INVOLVING COMBAT OPERATIONS)

9. General

a. **This category of support involves US forces actually conducting operations in support of the HN** (see Figure VI-4). This is different from providing equipment or training support in order to enhance the HN's ability to conduct its own operations. Direct support operations provide immediate assistance and are usually combined in a total FID effort with indirect operations.

b. Two types of direct support operations critical to supporting FID across all categories are CMO and PSYOP. Because these operations involve US forces in a direct operational role, they are discussed under direct support (not involving combat operations). Also included in this direct support discussion are military training to HN forces, logistic support, and intelligence and information sharing activities.

10. Civil-Military Operations

CMO span a very broad area in FID and include activities across the range of military operations. Using CMO to support military activities in a FID operation can enhance preventive measures, reconstructive efforts, and combat operations in support of an HN's IDAD program. This discussion is limited to those portions of CMO that most directly contribute to a commander's support of a FID operation.

For further information on CMO relationships to CAO, refer to JP 3-57, Civil-Military Operations.

a. **Civil Affairs Operations.** CAO enhance the relationship between military forces and civil authorities in areas where military forces are present. CAO, usually planned, directed, and conducted by CA personnel because of the complexities and demands for specialized capabilities involved in working within areas normally the responsibility of indigenous civil governments or authorities, enhance the conduct of CMO. **CAO are vital to theater FID operations in areas from planning to execution.** They are a valuable resource in planning and facilitating the conduct of various indirect, direct support (not involving combat operations), and combat operations in support of the overall FID effort. CAO also support the reconstitution of a viable and competent civil service and social infrastructures in areas of the HN that were previously ungoverned or under-governed or in the direct control of threat forces or shadow governments. CAO

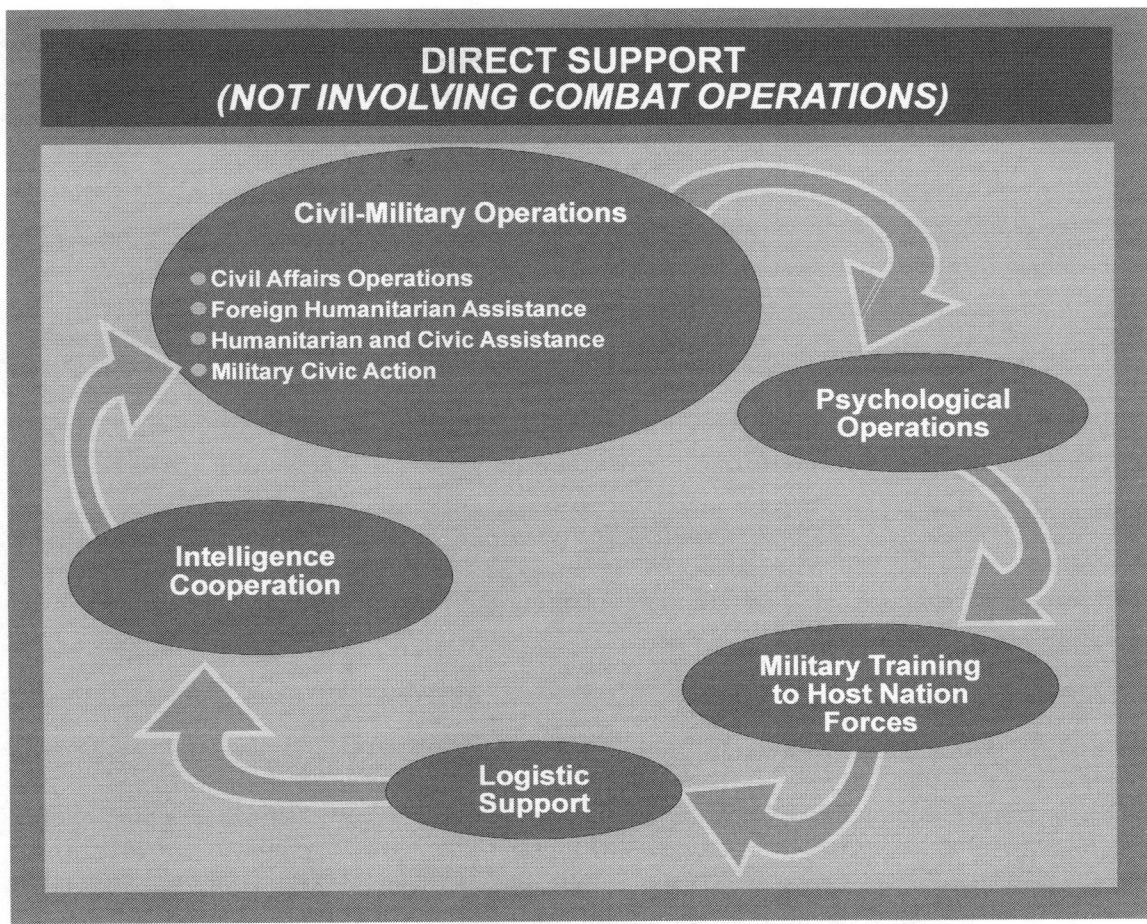

Figure VI-4. Direct Support (Not Involving Combat Operations)

can also assist the HN civilian government by providing civil administration assistance within its governmental structure.

(1) **Force Structure.** Each GCC is aligned with both an AC Army CA element and with an Army CA command, which are found only in the US Army Reserve and staffed with functional experts possessing a wide range of functional specialty areas.

(a) **Civil Affairs Liaison.** CA liaison personnel may be assigned or attached to a combatant command to augment the CA staff planning element. These personnel can be instrumental in assisting in the planning of military operations to support FID and incorporating FID operations into the overall theater strategy.

(b) **Civil Affairs Commands, Brigades, and Battalions**

1. US Army CA deploy as teams, detachments, or larger units in support of a JFC. US Army RC CA typically support CF requirements, while US Army AC CA are SOF and support SOF requirements. Accordingly, US Army RC CA are sourced through USJFCOM (via US Army Headquarters), whereas US Army AC CA are sourced through USSOCOM. CAO planners at United States Army Civil Affairs and

Psychological Operations Command (Airborne) and USSOCOM have the ability to determine the proper doctrinal structure or task organization to meet requirements.

$\underline{2}$. US Marine Corps CA deploy as teams, detachments, companies, or groups integral to a MAGTF or, when made available, in support of a JFC.

$\underline{3}$. Navy Civil Affairs. The maritime civil affairs group (MCAG) represents the US Navy's CA capabilities. MCAG has the capability to deploy maritime CA teams, maritime CA planners, and functional team specialists. MCAG was established to support the joint force maritime component commander, the Navy component commander and the GCC's SC strategy contained in the TCP by meeting Navy unique requirements in delivering CMO within the maritime domain. Three maritime-specific capabilities are available, in addition to the 14 found in the US Army:

$\underline{a}$. Port operations;

$\underline{b}$. Harbor and channel construction and maintenance; and

$\underline{c}$. Marine and fisheries resources.

(2) **Civil Affairs Capabilities.** CA units (both AC and RC) are regionally focused, possess varying levels of language capability, and country expertise. Only the RC have all six CA functional specialty areas: **governance** (public administration; public safety; environmental management); **rule of law** (judicial administration; corrections; public safety; law enforcement); **infrastructure** (public transportation; public works and utilities; public communications); **economic stability** (food and agriculture; economic development; civilian supply); **public health and welfare** (public health and cultural relations); and **public education and information** (public education; civil information services). Commanders should consider using their CA assets in the following roles to support the overall FID effort.

(a) Planning, supporting, and controlling other military operations in FID such as training assistance, FHA, MCA, HCA, and logistic support.

(b) Providing liaison to civilian authorities, IGOs, and NGOs.

(c) Facilitating the identification and procurement of civilian resources to support the mission.

(d) Supporting and conducting civil administration.

(e) The Coast Guard does not maintain CA. However, the Coast Guard can provide a variety of capabilities, assistance, equipment, and training in helping a country organize and establish a coast guard. Generally, maritime USCG forces have four principal missions: military operations and preparedness, law enforcement, maritime safety (including search and rescue), and enforcement of shipping and navigation laws.

The Coast Guard Model Maritime Service Code is a valuable reference for other nations to use for establishing a maritime force.

(3) **Civil Affairs Employment Considerations in FID.** The following are areas that commanders must consider when employing CA assets in planning, supporting, and executing FID operations.

(a) CA expertise must be incorporated in the planning as well as into the execution of military activities in support of FID operations.

(b) Successful FID operations hinge upon HN public support. Integrating CAO and PSYOP with FID operations can enhance that support. For specific humanitarian and indigenous religious leader liaison missions, chaplain support may offer mission credibility.

(c) The sovereignty of the HN must be maintained at all times. The perception that the US is running a puppet government is counter to the basic principles of FID. This is important to remember when providing civil administration assistance.

(d) HN self-sufficiency must be a goal of all CAO.

Additional information on CAO planning guidance, force apportionment, logistics, and joint forces C2 of CAO can be found in CJCSI 3110.12D, Civil Affairs Supplement to the Joint Strategic Capabilities Plan (U).

b. **Foreign Humanitarian Assistance. FHA programs are conducted to relieve or reduce the results of natural or man-made disasters or other endemic conditions.** FHA provided by US forces is generally limited in scope and duration. The assistance provided is designed to supplement or complement the efforts of the HN civil authorities or agencies that may have the primary responsibility for providing FHA. FHA may be planned into the GCC's military strategy to support FID as a component of the overall program to bolster the HN's IDAD capability. Often, however, FHA efforts are in response to unforeseen disaster situations. FHA efforts may also extend outside the FID umbrella. When FHA is provided to a nation that is experiencing lawlessness, subversion, or insurgency, these efforts must be considered as part of the FID effort. As such, all of the PSYOP and CMO/CAO considerations discussed earlier must be considered as the FHA programs are planned and executed.

(1) **FHA Missions and Assistance.** A single FHA operation may contain one or more FHA missions. Common missions include: relief missions, dislocated civilian support missions, security missions, technical assistance and support functions, and consequence management operations. Common examples of FHA that commanders may provide and/or restore are temporary shelter, food and water, medical assistance, transportation assistance, or other activities that provide basic services to the local populace. These services are often in response to a natural disaster such as an earthquake, a volcanic eruption, or a flood. In addition, FHA support may include

assistance to the populace of a nation ravaged by war, disease, or environmental catastrophes. Missions also could be conducted within the profile of consequence management.

(2) FHA Coordination and Control

(a) **DOS.** The US ambassador to the affected nation is responsible for declaring the occurrence of a disaster or emergency in a foreign country that requires US FHA support. This declaration is sent to the Office of US Foreign Disaster Assistance (OFDA) and DOS to begin possible USG assistance. **USAID, which is under the direct authority and foreign policy guidance of the Secretary of State, acts as the lead federal agency for US FHA.** USAID administers the President's authority to provide emergency relief and rehabilitation through OFDA. Should OFDA request that DOD conduct certain FHA operations, DOS would reimburse DOD for those operations. Appendix A, "Legal Considerations," covers the legal authorizations and restrictions for these operations.

(b) **DOD.** The USD(P) has the overall responsibility for developing military policy for FHA operations. The ASD(SO/LIC&IC) administers policy and statutory programs. Policy oversight is executed by the Deputy Assistant Secretary of Defense (Stability Operations). Program management and funding of these programs is the responsibility of DSCA. Much of FHA is provided through the excess property authorization under Title 10, USC, Section 2547, which permits the transfer of excess DOD property to authorized nations.

(c) **Chairman of the Joint Chiefs of Staff.** CJCS is responsible for recommending supported and supporting commands for FHA operations. The Joint Staff's Director for Operational Plans and Joint Force Development has the primary responsibility for concept review of OPLANs in support of FHA; the J-4 oversees

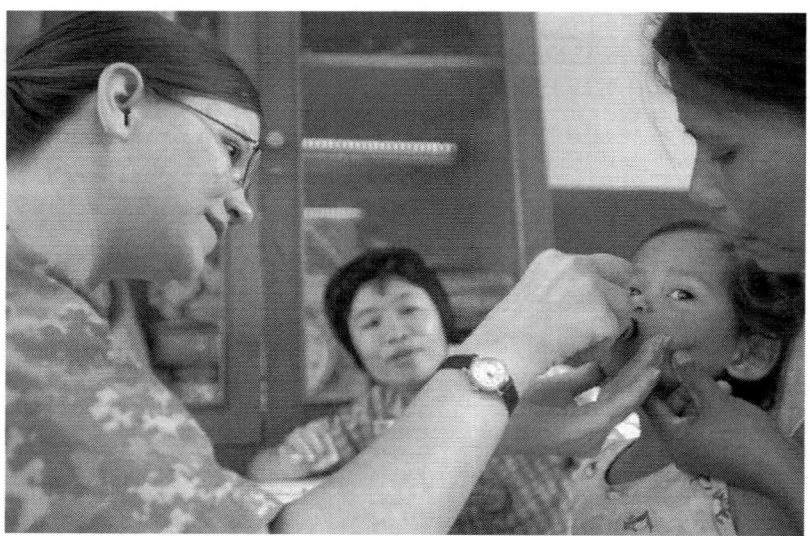

Medical assistance programs employ joint forces to promote nonmilitary objectives.

Service logistic support; and the J-3 will be involved when a military force is inserted into a foreign country as part of a US humanitarian response.

(d) **Geographic Combatant Commander.** The GCC considers FHA when formulating and establishing theater strategic objectives. Planning is conducted in accordance with JOPES. The supported GCC structures the force necessary to conduct and sustain FHA operations. In certain circumstances where coordination and approval lead times are not adequate, the GCC is authorized to commit the command's resources to provide immediate assistance.

(3) **FHA Employment Considerations.** Certain major points that CCDRs and other JFCs should consider when planning or executing FHA operations are listed in Figure VI-5.

For further information on FHA, refer to JP 3-29, Foreign Humanitarian Assistance.

c. **Humanitarian and Civic Assistance.** These HCA programs can be very valuable to the GCC's support of FID operations, while at the same time offering valuable training to US forces. It is important to understand the difference between HCA and FHA programs. FHA programs, as discussed above, focus on the use of DOD excess property, emergency transportation support, disaster relief, or other support as necessary to alleviate urgent needs in a host country caused by some type of disaster or catastrophe. HCA programs are specific programs authorized under Title 10, USC, Section 401 funding. **These programs are designed to provide assistance to the HN populace in conjunction with a military exercise.** These are usually planned well in advance and are usually not in response to disasters, although HCA activities have been executed following disasters. Specific activities for which HCA funds can be used include medical, dental, and veterinary care provided in areas of a country that are rural or are underserved by medical, dental, and veterinary professionals, respectively; construction of rudimentary surface transportation systems; well drilling and construction of basic sanitation facilities; rudimentary construction and repair of public facilities; and detection and clearance of landmines and other explosive remnants of war, including activities relating to the furnishing of education, training, and technical assistance with respect to the detection and clearance of landmines and other explosive remnants of war. **Assistance fulfills unit training requirements that create humanitarian benefit to the local populace.**

(1) **HCA Coordination and Control.** The coordination requirements for HCA projects are similar to those for FHA. HCA activities (other than *De Minimis* HCA, as defined in DOD Directive (DODD) 2205.2, *Humanitarian and Civic Assistance Provided in Conjunction with Military Operations*) conducted in a foreign country require the specific approval of the Secretary of State. Within DOD, the USD(P) oversees HCA, and the Deputy Assistant Secretary of Defense, Partnership Strategy and Stability Operations, acts as the program policy lead. GCCs develop a proposed annual execution plan for HCA activities within their AORs and execute HCA activities in conjunction with other military operations. Coordination with USAID and the country team is also very

FOREIGN HUMANITARIAN ASSISTANCE EMPLOYMENT CONSIDERATIONS

Constructing likely event joint operation plans. Theater joint operation plans should be established for likely events that may require a foreign humanitarian assistance (FHA) response. For example, if the combatant commander's area of responsibility covers a likely earthquake region, the command's response for an earthquake should be included in a joint operation plan to decrease response time and make execution easier and more effective.

Incorporating psychological operations and civil affairs operations into the FHA program. FHA programs that are not properly coordinated with local government officials, improperly assessed, or are misrepresented by opposition propaganda efforts may severely damage the theater foreign internal defense operation.

Coordinating all FHA activities with the country team. The US Agency for International Development and its subordinate element, the Office of Foreign Disaster Assistance, must be carefully consulted on disaster assistance. In addition, ensure that all proposed FHA efforts are coordinated with the Office of the Assistant Secretary of Defense (Special Operations/Low-Intensity Conflict and Interdependent Capabilities). The Department of State is the lead agent for this type of support and all FHA programs must be thoroughly coordinated to ensure an efficient national-level program.

Monitoring the legal authorizations carefully. These authorizations are subject to change and misinterpretation. A legal review must be conducted before planning FHA missions.

Figure VI-5. Foreign Humanitarian Assistance Employment Considerations

important to efficient HCA operations. The HCA program is a more decentralized program than FHA. This allows the GCC to have greater influence and to plan for a larger role in this area.

(a) In past years, Congress has established a ceiling on HCA authorizations. Funds are included in the GCC's budgets. This allocation is made based on national security priorities and guidance from SecDef.

(b) Valid HCA priorities within the AOR are identified and included in the FID planning process. These priorities then compete with those of other combatant commands for approval and funding. The resulting exercises and projects should then be managed as both FID and US training.

*Humanitarian and civic assistance as part of
foreign internal defense activities.*

(2) **HCA Employment Considerations.** The nature of HCA operations makes the employment considerations for this type of operation quite different than for FHA. The following are the key employment considerations for HCA.

(a) Plan for use of RC as well as AC forces. The medical, veterinary, dental, construction, and well drilling activities of HCA are well suited to skills found in the RC forces. Typical HCA missions allow these forces to get the realistic training they may not receive at their home stations or at other US training sites.

(b) Incorporate PSYOP and CA into HCA just as in FHA operations.

(c) Subject all HCA plans to close legal scrutiny. Like FHA operations, the legal aspects of all HCA operations must be understood.

(d) Plan adequate deployed FP measures. HCA operations are sometimes conducted in areas that are subject to unrest and internal instability. US forces may become targets of opposition forces' terrorist attacks. Security measures must be included in each operation.

(e) Establish the primary purpose of HCA missions as training for US forces. Incidental to this purpose are the benefits received by the civilian population. Nevertheless, every attempt should be made to ensure activities support HN capacity building.

(f) Conduct all HCA operations in support of the host civilian populace. No HCA projects/programs may be provided to HN military or paramilitary forces.

For further information on HCA, refer to JP 3-29, Foreign Humanitarian Assistance.

d. **Military Civic Action.** MCA programs offer the JFC a CMO opportunity to improve the HN infrastructure and the living conditions of the local populace, while enhancing the legitimacy of the HN government. **These programs use predominantly indigenous military forces at all levels in such fields as education, training, public works, agriculture, transportation, communications, health, sanitation, and other areas that contribute to the economic and social development of the nation.** These programs can have excellent long-term benefits for the HN by developing needed skills and by enhancing the legitimacy of the host government by showing the people that their government is capable of meeting the population's basic needs. MCA programs can also be helpful in gaining public acceptance of the military, which is especially important in situations requiring a clear, credible demonstration of improvement in host-military treatment of human rights. MCA is a tool that GCCs and subordinate JFCs should use, whenever possible, to bolster the overall FID plan.

For further information on MCA, refer to JP 3-57, Civil-Military Operations.

(1) **MCA Examples.** US forces may advise or assist the HN military in conducting the MCA mission. This assistance may occur in conjunction with SA training or as a GCC's separate initiative. In all cases, the actual mission must be performed by the HN military. Some of the most common MCA projects are in the area of construction.

(2) **MCA Coordination and Control.** Coordination for MCA missions is slightly less involved than for FHA and HCA missions. First, the US level of involvement is generally less than that required for other types of FID missions. Second, the program is essentially a US military to HN military project. As with all FID operations, however, the US ambassador and country team should be aware of all operations in their assigned country. If the US military support to MCA is provided through SA, normal SA coordination procedures apply, but if it is provided through a separate GCC's initiative using O&M funds, most of the coordination will be internal to the command.

(3) **MCA Employment Considerations.** Many of the same considerations apply when employing US military personnel in support of MCA as in supporting FHA and HCA. The essential difference is that in MCA, US personnel are limited to training and advisory roles. In addition to this general point, commanders should also consider the following employment guidelines when planning or executing MCA programs.

(a) Select projects that are simple and achievable and can be maintained by the HN. If the HN military is unable to accomplish the mission, confidence in the local government and military may be significantly damaged.

(b) HN forces will do the work required to accomplish the mission.

(c) Because of the nature of MCA missions, commanders will normally include CA, PSYOP, other SOF trainers, and combat support and combat service support elements to support MCA missions.

(d) Coordinate projects with the country team. The USAID representative should be consulted for assistance on any major MCA developmental project and should be informed of all MCA efforts.

11. Psychological Operations

a. **PSYOP supports the achievement of national objectives by influencing behaviors in select target foreign audiences.** During FID, PSYOP units design series that may include both PSYOP products and actions to convey to the TA US resolve, and the extent to which national objectives will be attained. The main objectives of PSYOP during FID are to build and maintain support for the host government while decreasing support for insurgents.

(1) **Target Groups and PSYOP Goals Within FID**

(a) **Insurgents.** To create dissension, disorganization, low morale, subversion, and defection within insurgent forces.

(b) **Civilian Populace.** To gain, preserve, and strengthen civilian support for the HN government and its IDAD program.

(c) **Military Forces.** To strengthen military support, with emphasis on building and maintaining the morale of HN forces.

(d) **Neutral Elements.** To gain the support of uncommitted groups inside and outside the HN.

(e) **External Hostile Powers.** To convince hostiles that the insurgency will fail.

(2) **PSYOP Activities**

(a) Improving popular support for the HN government.

(b) Discrediting the insurgent forces with neutral groups and the insurgents themselves.

(c) Projecting a favorable image of the HN government and the US.

(d) Supporting defector programs.

(e) Providing close and continuous support to CMO.

(f) Supporting HN programs that protect the population from insurgent activities.

(g) Strengthening HN support of programs that provide positive populace control and protection from insurgent activities.

(h) Informing the international community of US and HN intent and goodwill.

(i) Passing instructions to the HN populace.

(j) Developing HN PSYOP capabilities.

(3) **PSYOP Force Structure.** Similar to the CA force structure, over 85 percent of PSYOP assets are in the RC force. The Army has the preponderance of PSYOP assets. The US Air Force has a variety of assets capable of supporting PSYOP including a limited number of EC-130 COMMANDO SOLO aircraft. The EC-130J COMMANDO SOLO is a specially modified C-130 that conducts PSYOP and CA broadcasts. A typical mission consists of a single-ship orbit offset from the desired audience (either military or civilian personnel).

(4) **PSYOP Capabilities.** PSYOP personnel prepare an in-depth analysis of the target nation's social, political, religious, cultural, and economic environment as background for the development of the supporting PSYOP plan. Taking into consideration social groups, key communicators, and centers of gravity the PSYOP planner will recommend a theme (subject, topic, or line of persuasion used to achieve a psychological objective) to the GCC. The PSYOP planner recommends certain symbols that have a psychological impact on the TA. Once the commander chooses a theme and symbols, the tone and general parameters for much of the PSYOP to support the FID operation have been established, and all military operations should be evaluated against these parameters. Objectives, supporting objectives, and themes frame the program that will reach foreign TAs and reflect national and theater policy and strategy. **Approval of broad themes and messages is reserved by USG policy at OSD levels where the interagency process can address PSYOP products with a broad range of considerations.** In addition to establishing the psychological theme and symbols, the PSYOP element will project a favorable image of US actions, using all resources to channel the behavior of the TA so that it supports US objectives. Given these extensive capabilities, commanders should use their PSYOP assets to complement the FID plan in the following roles:

(a) Providing planning assistance for military support to FID. Planning tasks include identification of those military operations conducted primarily for their psychological effects and review of other military operations that have a psychological impact.

(b) Working with the military PAO and DOS PA personnel to build an extensive information effort to inform the local populace of US intentions in the FID effort and to strengthen the credibility of the HN government.

(c) Gathering information through PSYOP assessments of the local area that assist in determining FID requirements and MOEs.

(5) **PSYOP Employment Considerations in FID.** The following are areas that commanders must consider when employing PSYOP assets in support of FID operations.

(a) Accurate intelligence is imperative to successful PSYOP and FID. An inadequate analysis of the TA could result in the use of improper themes or symbols and damage the entire FID effort.

(b) Early integration of PSYOP to FID planning is imperative at all levels. PSYOP planning for FID requires an extensive knowledge of the TA vulnerabilities, social and cultural dynamics, and languages. Late integration of PSYOP to the FID operation may degrade the USG efforts to strengthening the HN credibility to the populace.

(c) PSYOP programs are audience driven; an analysis is required for each new TA and must be updated as attitudes and vulnerabilities change.

(d) PSYOP is a combat multiplier and should be used as any other capability. This use includes evaluation of targets through joint targeting procedures.

(e) Military PSYOP programs should be coordinated and synchronized with other USG information efforts.

b. **Peacetime PSYOP Programs.** GCCs may develop such programs, in coordination with the chiefs of US diplomatic missions, that plan, support, and provide for the conduct of PSYOP in support of US regional objectives, policies, interests, and theater military missions.

For further information on PSYOP, refer to JP 3-13.2, Psychological Operations.

12. Military Training to Host Nation Forces

a. The HN FID situation may intensify and increase the need for military training beyond that of indirect support. Direct support operations should provide more immediate benefit to the HN and may be used in conjunction with various types of SA indirect support training.

b. Increased emphasis on IDAD becomes important and training may focus on specific subversion, lawlessness, or insurgency problems encountered by the HN that may be beyond its capabilities to control.

c. **Security Force Assistance**

(1) The current strategic environment is seen as one of persistent conflict in which US national strategic objectives require unified action to achieve unity of effort. These synchronization, coordination, and integration activities of both governmental organizations and NGOs with military operations also require close relationships with US coalition partners, key friends, and allies. A key to achieving unity of effort is strengthening interagency coordination, working with international allies and partners, and reliance on strategic communication by the US and its international partners.

(a) Security sector reform supports unity of effort in FID by enabling allies, coalition members, and other nations to improve the way they provide safety, security, and justice.

(b) Within the context of security sector reform, the USG, including DOD, has long engaged in a range of activities to enhance the capacity and capability of partner nations by organizing, training, equipping, rebuilding and building, and advising and assisting FSF. See Figure VI-6. This is known as SFA.

(2) SFA encompasses joint force activities conducted within unified action to organize, train, equip, rebuild and build, and advise and assist FSF in support of an HN's efforts to plan and resource, generate, employ, transition, and sustain local, HN, or regional security forces and their supporting institutions. This includes activities from the ministry level to the tactical units, and the national security sector.

(a) SFA spans the range of military operations from military engagement, SC, and deterrence activities to crisis response and contingency operations, and if necessary, major operations and campaigns. It can include combat advisory and support activities not falling under SA. SFA does not include direct combat by US forces, as direct combat does not build the capability or capacity of FSF. Direct combat may assist the FSF in the same sense as a US unit assisting another US unit. This assistance, however, is in a generic sense, not in the sense of building capability or capacity and consequently is not SFA. SFA may be conducted in both permissive and uncertain security environments.

(b) Joint forces can conduct SFA unilaterally when necessary; however, when conducted within unified action, joint forces collaborate closely with interagency and multinational partners. See Figure VI-7.

FOREIGN SECURITY FORCES

Foreign security forces include but are not limited to the following:

- ■ Military forces

- ■ Police forces

- ■ Border police, coast guard, and customs officials

- ■ Paramilitary forces

- ■ Forces peculiar to specific nations, states, tribes, or ethnic groups

- ■ Prison, correctional, and penal services

- ■ Infrastructure protection forces

- ■ Governmental ministries or departments responsible for the above forces

Figure VI-6. Foreign Security Forces

(3) **Relationships Between FID and SFA**

(a) The basis for FID is established in US law; involves application of the instruments of US national power in support of a foreign nation confronted by internal threats; and focuses on the US efforts within unified action to enhance that nation's IDAD programs. FID military operations support the other instruments of national power through a wide variety of activities.

(b) As addressed in Chapter I, "Introduction," SFA is DOD's contribution to unified action to develop FSF capacity and capability from ministerial down to units of those forces. At operational and strategic levels, both SFA and FID focus on preparing FSF to combat lawlessness, subversion, insurgency, and terrorism from internal threats; however, SFA also prepares FSF to defend against external threats and to perform as part of an international coalition as well.

(c) FID and SFA are similar at the tactical level where advisory skills are applicable to both.

SECURITY FORCE ASSISTANCE

Security force assistance (SFA) is unified action to generate, employ, and sustain local, host nation, or regional security forces in support of a legitimate authority.

- SFA actions include joint, interagency, intergovernmental, multinational, nongovernmental, and private company cooperative efforts to ensure and support unity of effort / unity of purpose.

- US forces assist foreign security forces (FSF) development and operation across the range of military operations -- combating internal threats such as insurgency, subversion, and lawlessness, i.e., foreign internal defense.

- To be successful, SFA must be based on solid, continuing assessment and include the organizing, training, equipping, rebuilding, and advising of the forces involved. It is critical to develop the institutional infrastructure to sustain SFA gains.

- The resulting FSF must possess the *capability* to accomplish the variety of required missions, with sufficient *capacity* to be successful and with the ability to *sustain* themselves as long as required.

Figure VI-7. Security Force Assistance

(d) Both FID and SFA are subsets of SC. Neither FID nor SFA are subsets of each other.

(4) CF and SOF conducting SFA will find that initial and continuous assessment of the HN security forces is essential for successful advisory efforts. Particular attention is required to the organizing, training, equipping, rebuilding, and advising of the forces involved. Continuous assessment is essential throughout advisory operations. A comprehensive assessment will help advisors develop program objectives and milestones and a baseline to measure FSF progress and success.

(a) **Organizing.** SFA includes organizing institutions and units, which can range from standing up a ministry to improving the organization of the smallest maneuver unit. Building capability and capacity in this area includes personnel, logistics, and intelligence, and their support infrastructure. Developing HN tactical capabilities alone is inadequate; strategic and operational capabilities must be developed as well. HN organization and units should reflect their own unique requirements, interests, and capabilities—they should not simply mirror existing external institutions.

(b) **Training.** Training occurs in training centers, academies, and units. Training includes a broad range of subject matter to include security forces responding to civilian oversight and control.

(c) **Equipping.** Equipping is accomplished through traditional SA, foreign support, and donations. The equipment must be appropriate for the physical environment of the region and the HN's ability to operate and sustain it.

(d) **Rebuilding and Building.** In many cases, particularly after major combat operations, it may be necessary to rebuild existing or build new infrastructure to support FSF. This includes facilities and materiel but may also include other infrastructure such as C2 systems and transportation networks.

(e) **Advising and Assisting.** Advising HN units and institutions is essential to the ultimate success of SFA. This benefits both the state and the supporting external organizations. To successfully accomplish the SFA mission, advising requires specially selected and trained personnel.

(5) **The SFA Imperatives.** Successful SFA operations require planning and execution consistent with the following imperatives:

(a) **Understand the Operational Environment.** This includes an awareness of the relationships between the players within the unified action framework, the HN population, and opposing threats.

(b) **Provide Effective Leadership.** Both coalition and HN leadership must fully comprehend the operational environment and be prepared, engaged, and supportive in order for the SFA effort to succeed.

(c) **Build Legitimacy.** The ultimate goal of SFA is to develop security forces that contribute to the legitimate governance of the HN population.

(d) **Manage Information.** This encompasses the collection, analysis, management, application, and preparation of information.

(e) **Ensure Unity of Effort/Unity of Purpose.** The command relationships must be clearly delineated and understood. Supported and supporting relationships will change over time.

(f) **Sustain the Effort.** This includes two major efforts: the ability of the US/coalition to sustain the SFA effort throughout the campaign, and the ability of the HN security forces to ultimately sustain their operations independently.

For additional information on each of the six SFA imperatives listed above, refer to Commander's Handbook for Security Force Assistance, 14 July 2008.

(6) **Plan and Resource.** The plan and resource SFA activity begins as commanders understand the operational environment and determine the requirements of FSF. It also ensures that the US provides SFA that achieves the objectives and end state of both the HN and US. The HN and US should then analyze the resource requirements and efforts so that developing FSF have sufficient and appropriate resources.

(7) USSOCOM is the designated joint proponent for SFA, with responsibility to lead the collaborative development, coordination, and integration of the SFA capability across DOD. This includes development of SFA in joint doctrine; training and education for individuals and units; joint capabilities; joint mission essential task lists; and identification of critical individual skills, training, and experience. Additionally, in collaboration with the Joint Staff and USJFCOM, and in coordination with the Services and GCCs, USSOCOM is tasked with developing global joint sourcing solutions that recommend the most appropriate forces (CF and/or SOF) for validated SFA requirements referred to the global force management process.

13. Logistic Support

a. Logistic support as discussed here does not include activities authorized under SA. **Logistic support operations are limited by US law without an ACSA. Such support usually consists of transportation or limited maintenance support, although an ACSA can allow additional support in areas beyond those.** However, ACSAs permit the reimbursable exchange of logistics support, supplies, and services with HN personnel. The existence, potential application, and limitations of an ACSA with the HN's military forces should be considered when planning.

b. In some cases, the President or SecDef may direct a show of force exercise to demonstrate support for the HN and to provide the vehicle for provision of logistic support. An example of such a show of force was Operation GOLDEN PHEASANT conducted in support of the Government of Honduras in 1987. In this case the internal threat was less severe than the combined over the border threat of an interventionist Sandinista regime in Nicaragua.

c. Logistic support is integrated into the overall theater FID plan. This is even more important if the supported nation is involved in an active conflict.

d. The following are major employment considerations that should be considered when providing logistic support as part of the theater FID effort.

(1) Develop definitive ROE and FP measures.

(2) Educate all members of the command on permissible activities in providing the logistic support mission.

(3) Build a logistics assessment file on logistic resources available in country. This database should include information of local supply availability, warehousing and maintenance facilities, transportation assets, line of communications (LOC), and labor force availability.

(4) Tailor the proper types of equipment maintenance and training sustainability packages to the needs of the HN.

(5) Consider utilizing a sea base to provide logistics support, if naval assets are available and if geographically supportable.

14. Intelligence and Information Sharing

a. Information cooperation is enabled by an information sharing environment that fully integrates joint, multinational, and interagency partners in a collaborative enterprise. An active intelligence liaison should be ongoing among the HN, country team, and CCDR's intelligence staff, thus establishing the basis for any intelligence and communications sharing. CI elements can provide this support with HN military CI elements, security service, and police forces when deployed in support of FID operations. Other national-level baseline intelligence support (e.g., geospatial intelligence [GEOINT]) may be provided via established or ad hoc memorandums of agreement or through national-level liaison teams as required. During the GCC's assessment of the operational area, the HN intelligence and communications capabilities should be evaluated. Based on this evaluation, the GCC is in a good position to provide or recommend approval of intelligence or communications assistance. **The sharing of US intelligence is a sensitive area that must be evaluated based on the circumstances of each situation.** Cooperative intelligence liaisons between the US and HN are vital; however, disclosure of classified information to the HN or other multinational FID forces must be authorized. Generally, assistance may be provided in terms of evaluation, training, limited information exchange, and equipment support.

b. Any intelligence assistance must be coordinated with the country team intelligence assets to benefit from operational and tactical capabilities.

c. The initial focus of assistance in this area will be to evaluate HN intelligence and communications architecture. Based on this evaluation, the GCC will be able to determine the HN's requirements.

d. Examine the intelligence process as it applies to the current situation. The needs of the HN as well as their technical expertise and equipment must be considered when evaluating their systems. The HN intelligence and communications systems must reflect the HN's environment and threat.

e. Following the evaluation, a determination must be made as to how the US FID operation may assist. Any intelligence sharing must be evaluated against US national security interests and be both coordinated and approved at the national level.

f. Equipment deficiencies should be identified in the assessment. US assistance in equipment normally will be provided through the SA process.

g. Training support for intelligence operations, which is indirect support, will also normally be conducted under SA. Some limited informal training benefits may also be provided during exchange programs and daily interface with HN military intelligence and communications assets.

h. **Employment Considerations.** The following items summarize the major considerations that commanders and planners must be aware of as they conduct intelligence cooperation activities in support of the FID operation.

(1) If the HN is not capable of performing intelligence and CI missions effectively upon the commitment of US FID forces, US intelligence and CI elements should be deployed to accomplish these missions. The effectiveness of CI assets is significantly increased with early introduction. Additionally, engaging existent CI resources through FPD and FAO should be considered in the planning stages to maximize CI resources.

(2) Direct most intelligence and information efforts toward creating a self-sufficient HN capability. US assistance that creates a long-term reliance on US capabilities may damage the overall HN intelligence and communications system.

(3) Scrutinize any training assistance to ensure that it is provided within legal authorizations and ensure that information or processes are not revealed without authorization.

(4) Tailor assistance to the level of the threat, equipment, and technology within the HN.

(5) Initial and limited use of US ISR technology may benefit the HN but must be balanced with the more lasting solution of realistically upgrading the HN's ISR capability in a manner it can sustain. HN intelligence professionals must understand the temporary nature of such US support.

(6) Because FID inherently involves extensive work with HN personnel, special attention should be given to releasability considerations of intelligence products, as well as available bandwidth and other potential interoperability concerns.

SECTION D. COMBAT OPERATIONS

15. General

a. US participation in combat operations as part of a FID effort requires Presidential authority.

b. In some cases, the Armed Forces of the US may be required to conduct COIN, CT, CD, or other sustained operations directly in the place of HN forces, particularly if HN security force capacity is still being developed. In other cases, the Armed Forces of the US may support HN forces conducting such operations by directly participating in combat operations.

16. Considerations for United States Combat Operations

a. This section discusses areas at the operational and strategic levels of war that should be considered when conducting combat operations in support of an HN's IDAD program. Many of the considerations discussed in the other two categories of FID remain important in tactical operations. The most notable of these involve the coordinated use of PSYOP and CAO as well as coordination with other USG agencies operating within the HN. The following are areas that the JFC also should consider when employing combat forces in support of FID.

b. **HN IDAD Organization.** Maintain close coordination with the elements of the government responsible for HN IDAD efforts. **If a nation has reached a point in its internal affairs that it requires combat support from the US, it should have already developed a comprehensive IDAD strategy.** The organization to effect this strategy will vary among nations. It may simply be the normal organization of the executive branch of the HN government. **The important point is that an organization should exist to pull together all instruments of power to defeat the source of internal instability.** JFCs must be involved in this coordination and control process. Chapter II, "Internal Defense and Development," provides a detailed explanation of an IDAD strategy as well as a sample IDAD organization. US commanders supporting an IDAD program must be integrated into the organizational structure that controls the program.

c. **Tiers of Forces.** Historically, COIN and stability efforts are led initially by SOF (principally SF, PSYOP, and CA). If HN capabilities are not sufficiently improved and MOEs are not attained, CF may be required. This is particularly true if the HN has one or more porous borders that internal threats operate across and even more likely if a bordering nation-state is providing support to the internal threat.

d. **Transition Points.** Establish transition points at which combat operations are to be returned to the HN forces. This process establishes fixed milestones (not time dependent) that provide indicators of success of the HN IDAD program.

e. **Joint, Interagency, and Multinational Focus.** Combat operations supporting FID will normally be joint, and may also include multinational operations involving the HN, the US, and other multinational forces. Interagency coordination should also be anticipated, given the inherent nature of FID operations.

f. **US Combat Operations.** A priority for US combat operations should be to identify and integrate logistics, intelligence, and other combat support means. When tactically feasible, actual combat operations and support for them should be done by HN forces, thus increasing the legitimacy of the HN government and reducing the dependency on US forces. US forces will conduct combat operations only when directed by legal authority to stabilize the situation and to give the local government and HN military forces time to regain the initiative. In most cases, the objective of US operations will be FP rather than to focus on destruction of adversary forces. Gaining the strategic initiative is the responsibility of the HN. Commanders must evaluate all operations to ensure that they do not create the impression that the US is executing a war for a nation that has neither the will nor the public support to defeat internal threats.

g. **Human Rights Considerations.** Strict adherence to respect for human rights must be maintained. This includes US forces as well as forces from the HN and other participating multinational forces. Repression and abuses of the local population by the legitimate government will reduce the credibility and popular support for the HN government and also may cause the President to consider withdrawing US support; therefore, commanders must consistently reinforce human rights policies. In many FID combat situations, the moral high ground may be just as important as the tactical high ground.

h. **Rules of Engagement.** Judicious and prudent ROE are absolutely required in combat operations in FID. A balance between FP and danger to innocent civilians as well as damage to nonmilitary areas must be reached. Each individual must be trained in order to prevent unnecessary destruction or loss of civilian life. Commanders must closely monitor this situation and provide subordinate commanders with clear and enforceable ROE as well as the flexibility to modify these ROE as the situation changes.

i. **Indiscriminate Use of Force.** Indiscriminate use of force is not authorized. Force may be employed only in accordance with US legal, policy, and applicable ROE standards.

j. **Intelligence.** The US joint intelligence network must be tied into the country team, the local HN military, paramilitary, and police intelligence capabilities, as well as the intelligence assets of other nations participating in the operation. Deployed military CI elements can provide this liaison with local HN military CI and security and police services in their areas of operations. In addition, social, economic, and political information must be current to allow the commander to become aware of changes in the operational environment that might require a change in tactics. Appendix B, "Joint Intelligence Preparation of the Operational Environment to Support Foreign Internal Defense," provides detail on the type of information necessary for a thorough evaluation of the area of operation. The nature of the required information places a greater emphasis on HUMINT efforts than on technical collection capabilities.

k. **FID Integration With Other Activities.** The initiation of hostilities does not mean that other FID operations will be suspended. In fact, PSYOP, CMO, SA, FHA, intelligence, and logistic support are all likely to increase dramatically. The FID planning imperatives of taking the long-term approach, tailoring support to HN needs, and the HN bearing IDAD responsibility remain important throughout both combat and noncombat operations.

17. Command and Control

The C2 relationships established for the combat operation may be modified based on the political, social, and military environment of the area. In general, the following C2 recommendations should be considered when conducting FID combat operations.

a. **The HN government and security forces must remain in the forefront.** The HN security forces must establish strategic policy and objectives, and a single multinational headquarters should be established to control combat operations.

b. **The chain of command from the President to the lowest US commander in the field remains inviolate.** The President retains command authority over US forces. It is sometimes prudent or advantageous to place appropriate US forces under the OPCON of a foreign commander to achieve specified military objectives. In making that determination, the President carefully considers such factors as the mission, size of the proposed US force, risks involved, anticipated duration, and ROE.

For further information on C2, refer to JP 1, Doctrine for the Armed Forces of the United States.

18. Sustainment

As with any operation, sustainment of US forces is essential to success. Sustainment of combat operations in FID is similar to sustainment for other types of operations. The political sensitivities and concern for HN legitimacy and minimum US presence do, however, change the complexion of sustainment operations in FID. The general principles that should be considered in planning and executing sustainment of combat operations in FID are:

a. **Maximum Use of HN Capabilities.** This includes routine services, supplies, facilities, and transportation. This approach reduces US overhead and the number of US personnel required in the HN.

b. **Maximum use of existing facilities such as ports, airfields, and communications sites.**

c. **Minimum Handling of Supplies.** For short duration operations (90 days or less), support will be provided through existing organic support packages, through air LOCs, or through the use of logistics flowing through a sea base.

d. **Medical Self-Sufficiency.** Many areas of the world where the US is likely to conduct FID do not have adequate medical capabilities. Since commanders cannot rely on local capabilities, they must plan for self-sufficient HSS for combat operations in FID. At a minimum adequate hospitalization, medical logistics resupply, patient movement, and preventive medicine must be established to support these operations.

e. **Optimum use of mobile maintenance capabilities that stress repair as far forward as possible.** Equipment evacuation for repair should be kept to a minimum.

f. **Routine use of both intertheater and intratheater airlift and sealift to deliver supplies.**

SECTION E. TRANSITION AND REDEPLOYMENT

19. General

a. Redeployment of units conducting FID operations does not typically indicate the end of all FID operations in the HN. Rather, in long-term FID operations as security and other conditions improve and internal threats become manageable for HN personnel, direct military-to-military activities by units will continue, but these activities may become more intermittent with gaps between regular exercises and exchanges. In ongoing FID operations, continuous coverage by US units generally involves mission handoff from one unit to its replacement. Redeployment, if conducted haphazardly or prematurely, can set FID operations back substantially.

b. Commonalities exist between redeployments that involve direct handoff and redeployments that involve intermittent deployments. In the latter case, the possibility always exists that situations during routine activities and military-to-military contact will arise causing handoff to a relieving force. Typically, this will involve the original unit or select members of it extending their presence in-country to provide continuity or to stay in place as part of a more robust force. Redeployment may also involve a transition from DOD execution of programs to DOS or OGA execution. In both immediate mission handoff and intermittent FID operations, capturing lessons learned in thorough post-mission debriefings is essential to continue to build institutional FID knowledge and refine FID doctrine and training.

20. Termination of Operations

The nature of the termination will shape the futures of the HN and regional countries. It is essential to understand that termination of operations is a vital link between national security strategy, national defense strategy, national military strategy, and the national strategic end state. A poorly conducted termination of FID operations can have a long-term impact on USG relations with the HN, the region, and, potentially, in more than one region. Some level of operations normally will continue well after intensive FID support has ended. The possibility of an extended presence by US military forces to assist FID operations should be considered during the initial planning and recommendation for execution.

21. Termination Approaches

a. There are three approaches for achieving national strategic objectives by military force. The first is to force an **imposed settlement** by the threat of or actual occupation of an enemy's land, resources, or people. Supporting the threat of actual occupation may be accomplished by the selective destruction of critical functions or assets, such as C2, infrastructure, or making the adversary unable to resist the imposition of US will. In FID, this approach typically is only taken with intransigent internal threats and the approach differs from other operations in that it still involves a preponderance of HN effort in any imposed settlement.

b. The second approach seeks a **negotiated settlement** through coordinated political, diplomatic, military, and economic actions, which convince an adversary that to yield will be less painful than continued resistance. In FID, military power alone will rarely compel an internal threat to consider a negotiated conclusion. Rather, military success in providing security to the HN populace coupled with the other functions of the HN IDAD program may induce an internal threat to negotiate under terms acceptable to the HN government. Negotiating an advantageous conclusion to operations requires time, power, and the demonstrated will to use both. However, some internal threats by their nature may not be viable candidates for negotiation. In addition to imposed and negotiated termination, there may be an armistice or truce, which is a negotiated intermission in operations, not a peace. In effect, it is a device to buy time pending negotiation of a permanent settlement or resumption of operations. The efficacy of an armistice or truce must be weighed against the potential damage done by legitimizing an internal threat.

c. The third approach for achieving national security objectives in relation to the irregular challenges posed by non-state actors is an **indirect approach** that erodes an adversary's power, influence, and will; undermines the credibility and legitimacy of its political authority; and undermines an internal threat's influence and control over and support by the indigenous population. This approach is necessary with an internal threat unwilling to enter into discussion.

22. National Strategic End State

a. The first and primary political task regarding termination of intensive FID operations is to determine an achievable national strategic end state based on clear national strategic objectives. For specific situations that require the employment of military capabilities (particularly for anticipated major operations), the President and SecDef typically establish a set of national strategic objectives. Achieving these objectives is necessary to attain the national strategic end state—the broadly expressed diplomatic, informational, military, and economic conditions that should exist after the conclusion of a campaign or operation. In FID, this is determined with the HN civilian leadership to ensure a clearly defined national strategic end state that is mutually beneficial. Specified standards are approved by the President or SecDef that must be met before a FID operation can be concluded or transitioned to a less intensive level of support.

b. Commanders clarify their desired end state for training programs early. The characteristics of effective HN security forces include flexible, proficient, self-sustained, well led, professional forces which are integrated into society. The well-trained HN security forces should:

(1) Provide reasonable levels of security from external threats while not threatening regional security.

(2) Provide reasonable levels of internal security without infringing upon the populace's civil liberties or posing a coup threat.

(3) Be founded upon the rule of law.

(4) Be sustainable by the HN after US and multinational forces depart.

23. Military Considerations

In its strategic context, military success is measured in the attainment of military objectives supporting the national strategic end state and associated termination criteria. Termination criteria for a negotiated settlement will differ significantly from those of an imposed settlement. Military strategic advice to USG and HN leadership regarding termination criteria should be reviewed for military feasibility and acceptability, as well as estimates of the time, costs, and military forces required to reach the criteria. An essential consideration is ensuring that the longer-term stabilization and the enabling of civil authority needed to achieve national strategic objectives continue upon the conclusion of sustained operations. Premature reduction of FID support can trigger a rapid and dramatic upsurge in internal threat activity, strength, and political viability. Proper use of the informational instrument of national power mitigates the possibility of any vestigial internal threat elements characterizing a reduction in military commitment to a US or HN strategic or tactical reversal.

24. Mission Handoff Procedures

a. During long-term continuous FID operations, commanders may elect to replace teams for a variety of reasons. Time is not the only governing factor. Changes in the HN operational environment may require reshaping force packages as situations change for better or worse. In addition, internal administrative concerns might prompt or support a commander's decision to rotate teams or units; for example, new equipment may be fielded to an incoming unit that the outgoing unit lacks. Regardless of reason, mission handoff is necessary and is defined as the process of passing an ongoing mission from one unit to another with no discernible loss of continuity.

b. The overall authority for the handoff and assumption of command lies with the commander ordering the change. The authority for determining the handoff process lies with the incoming commander since he will assume responsibility for the mission. This changeover process may affect the conditions under which the mission will continue.

c. The outgoing commander advises the incoming commander on the tentative handoff process and the assumption of the mission directly or through a liaison. If this advice conflicts with the mission statement or the incoming commander's desires and the conflict cannot be resolved with the authority established for the incoming commander, the commander ordering the relief resolves the issue.

d. As a rule, the commander ordering the change does not automatically place the outgoing unit under the incoming unit's control during the changeover process. Although this procedure would present a clear and easily defined solution to establishing the incoming commander's authority, it is not the most effective control for US forces should hostile contact occur during the process.

e. If the incoming US unit or the HN unit it advises is in direct-fire contact with insurgents or another internal threat during the handoff, the unit immediately notifies the higher headquarters ordering the exchange. If the incoming unit commander has not assumed responsibility, the unit immediately comes under control of the outgoing unit and is absorbed into that unit position. The outgoing unit commander and his or her HN counterpart will control the battle. If the outgoing unit commander has passed responsibility to the incoming unit commander, the outgoing unit comes under the OPCON of the incoming unit, and the HN unit coordinates its movements with the new unit. Units in advisory or combat support roles should follow these same procedures.

25. Considerations

Although the considerations listed below are intended primarily for a direct handoff between units, they also apply when handoff is made to an OGA. In addition, the considerations should be taken into account in a mature FID operation where there may be lag time between deployments. In this latter case, preparing an analysis of the considerations listed below will aid the incoming commander on the next iteration. The incoming and outgoing commanders or OGA lead representative should consider the following:

a. **Mission.** The incoming commander must make a detailed study of the unit's mission statement and understand the present mission tasks and the implied mission tasks. The mission may also require a unit with additional skill sets, such as specialized intelligence capabilities, near real time connectivity, CA functional specialists, or complex media production ability. Knowing the mission, commander's concept of the mission, commander's critical information requirements, PIRs, and IRs will help him or her understand the mission. After a complete in-depth study of the operational area, the incoming unit commander should complete the handoff in a manner that allows for continued, uninterrupted mission accomplishment. The changeover must not allow any adversary to gain operational advantages.

b. **Operational Environment.** The in-country unit provides continuous information updates to the incoming commander. PIRs and IRs were established for the original mission along with operational, strategic, and tactical information. The incoming unit must become familiar with the ongoing PIRs and IRs, and the upcoming mission PIRs and IRs.

c. **Adversary Composition.** The incoming unit commander must have the latest available intelligence on all internal threats that affect the mission. This intelligence includes comprehensive data on terrorists and terrorist-related incidents, criminal activity, and specific environmental threats over the previous several months. In addition to the normal intelligence provided to the incoming unit commander on a regular basis, the situation may call for a liaison from the outgoing unit. OPSEC is critical to prevent the enemy from discovering the impending relief and then exploiting the fluidity of the change and the concentration of US forces.

d. **Friendly United States/Multinational Forces.** To the incoming unit, learning about the friendly forces is as important as knowing the enemy situation. The unit must be familiar with the C2 structure it will deal with on a daily basis. The incoming unit must know all friendly units in adjacent operational environments and be aware of the capabilities of their mission support base in addition to other operations, units, and their capabilities. If US combat support units are to be relieved, their relief should occur after the relief of the units they support.

e. **Host Nation Forces.** The incoming unit plans and prepares for a quick and frictionless transition in counterpart relations. However, potential or anticipated friction between the HN unit and the incoming unit may cause the relief to take place more slowly than desired. Therefore, the incoming and outgoing units need a period of overlap to allow for in-country, face-to-face contact with their counterparts before the mission handoff. If possible, the incoming unit members should receive biographical data on their counterparts, to include photographs prior to deployment. This information allows unit members to become familiar with their counterparts and may aid to determine which advisor techniques need more emphasis. Mission execution must continue within the capabilities of the incoming unit, the HN unit, and the available supporting assets.

f. **Civilian Populace.** All incoming units must conduct an in-depth area study, giving close attention to local problems. General demographic data may be available from sister units that can be expanded upon for unit-specific needs. Popular support for US activities taking place within the operational environment may directly influence changes in the mission statement. The outgoing unit must provide this critical information and describe in detail all completed civic action projects and those that are underway. The incoming unit must understand the functioning of the HN government and the status of any international civilian or government agencies involved in or influencing the situation in its operational environment.

g. **Terrain and Weather.** Some handoff operations may require select SOF units such as SF detachments, CA teams, or tactical PSYOP teams to move by foot or by animal mounts into and out of the area of operations. In such instances, the outgoing unit plans and reconnoiters the routes used for infiltrating the incoming unit and those used for its exfiltration. These routes must provide the best possible cover and concealment.

If possible, the units make this exchange during darkness or inclement weather. SOF units must consider significant terrain or weather features that may impede movement. Limitations of media coverage or difficulties to civic action projects because of these features are two common examples. In addition, weather conditions and significant elevations can greatly affect air operations. These factors can critically affect resupply and health service and supply (notably medical evacuation procedures) as well.

h. **Time.** The depth and dispersion of units and the number of operations conducted will determine the time required to exchange units. Ideally, there is an overlap period to allow the incoming unit to become familiar with the operational environment and to establish rapport between the incoming unit personnel and their HN counterparts. However, the handoff operation must take place as quickly as possible. The longer the operation takes, the more personnel in the operational environment become a vulnerable and lucrative target. A quickly executed relief will reduce the time available to the enemy to strike before the incoming unit has time to consolidate its position. The incoming unit should not sacrifice continued and uninterrupted execution of ongoing operations for speed. The incoming unit needs to have enough time to observe training techniques and procedures and to conduct debriefing on lessons learned.

i. **Other Government Agencies.** For the unit being relieved of a function by an OGA handoff, procedures will typically take more time and entail more complex coordination. However, the other areas of consideration still apply and in fact may be a greater issue for an OGA. Outgoing units that have past, present, or future projects planned with OGAs must provide for the transfer of these projects to new responsible agents in the incoming unit. Outgoing unit personnel need sufficient time to put incoming unit personnel in contact with OGA counterparts. In addition, the outgoing unit should brief the incoming unit on any OGA programs affecting FID operations.

j. **Continued Joint Forces Involvement.** The constant and unbroken presence of SOF or CF in FID operations is not a foregone conclusion. In FID operations, gaps in deployments of SOF, CF, or both types of units may be unavoidable. In addition, limited CF or SOF units or advisors may be present on an ongoing basis. In these instances, the SOF or CF with a constant presence must maintain continuity and brief sister-SOF units or conventional units on aspects of the operational environment. Various operational concerns can affect what sort of unit relieves another, for instance, a conventional force after completing the basic training of an elite but previously untrained HN unit might handoff to an SOF unit to complete the HN unit's advanced training.

26. Post Mission Debriefing Procedures

a. The unit commander conducts a debriefing that provides an overview of the mission and all relevant informational subsets. The debrief should begin with the updated area study, and continue with other relevant issues. The range of topics can include military geography; political parties; military forces; insurgents; security forces; underground, lethal and nonlethal targets and TAs; ongoing CAO; logistics; health service and supply issues; and ongoing joint, interagency, intergovernmental, and multinational projects or operations. Figure VI-8 depicts a post mission debriefing guide.

b. **Documentation.** As FID operations are executed and the joint force rotates, it is critical to document lessons learned to allow the commander to modify the FID operation to fit the special circumstances and environment. Debriefs by individual units of the joint force ensure internal continuity. Relevant portions of these debriefs are consolidated into a single joint force after-action report. Therefore, comprehensive after-action reviews and reports focusing on the specifics of the FID operations should be conducted to gather this information as soon as possible after mission execution.

For further information on documenting lessons learned, refer to CJCSI 3150.25D, Joint Lessons Learned Program (JLLP).

POST MISSION DEBRIEFING GUIDE

MISSION
- Brief statement of mission by joint force commander.

EXECUTION
- Brief statement of the concept of operation developed before the deployment.
- Statement of method of operation accomplished during the operation, to include deployment, routes, activity in host nation (HN) areas, and redeployment.
- Uniforms and equipment used.
- Weapons, demolitions, and ammunition used and results.
- Communications and media equipment used and results.
  - Organic.
  - HN force.
  - Contract.
- Casualties (friendly and enemy) sustained and disposition of bodies of those killed in action.
- Friendly contacts established, to include descriptions, locations, circumstances, and results.

MILITARY GEOGRAPHY
- Geographic name, Universal Transverse Mercator or geographic coordinates, and locations.
- Boundaries (north, south, east, and west).
- Distance and direction to nearest major cultural feature.
- Terrain.
  - What type of terrain is dominant in this area?
  - What natural and cultivated vegetation is present in the area?
  - What is the density and disposition of natural vegetation?
  - What is the approximate degree of slope?
- What natural obstacles to movement were observed, and what are their locations?
- What natural or man-made obstacles to media or humanitarian distribution are there?
- What natural or man-made drainage features are in the area?
  - Direction of flow.
  - Speed.
  - Depth.
  - Type of bed.
- What is the physical layout of rural and urban settlements?
- What is the layout of various houses within the area?
- What is the description of any potential landing zones or drop zones?
- What is the description of any beach landing sites, if applicable?
- What are the descriptions of any areas suitable for cache sites, and what are their locations?
- People.
  - What major ethnic groups or tribes populate each area?
  - What was (or is) their attitude toward other ethnic groups or tribes in the area?
  - What are the principal religion(s) of the area, and how are they practiced?
    - o Main, secondary, etc. status.
    - o Influence on people.
    - o Influence or control on political or judicial processes.
    - o Religious prayer times, regular observed days, and holidays.
    - o Constraints, laws, and taboos.
    - o Conflicts in or between religions, denominations, or sects.
    - o Religious themes, symbology, and allegory or folklore.
    - o View on conflict and martyrdom.
  - What is the description of the average citizen of the area (height, weight, hair color, characteristics)?
    - o Is there a physically differentiated minority?
    - o Is there a minority differentiated by other visual cues such as dress or hairstyle?
  - What type clothing, footwear, ornaments, and jewelry do they wear?
  - Is symbolism or status attached to certain items of jewelry or ornaments?
  - What are the local traditions, customs, and practices?
    - o Between males and females?
    - o Between young and old?

Figure VI-8. Post Mission Debriefing Guide

MILITARY GEOGRAPHY (continued)

- o Toward marriage, birth, and death?
- o Between the populace and local officials?
- What is the ordinary diet of the people?
  - o Self-imposed restrictions.
  - o Chronic dietary deficiencies.
  - o Cyclic, seasonal, or localized deficiencies.
- What was the attitude of the populace toward US and HN forces?
  - o Friendly target groups or specific target audiences (TAs).
  - o Neutral or uncommitted target groups or specific TAs.
  - o Hostile target groups or specific TAs.
  - o Specific behavioral changes noted.
  - o Anecdotal occurrences or spontaneous events during current deployment.
- What was the general feeling and attitude of the populace and the HN troops toward the government and leaders, government policies, and general conditions within the country?
- How did the populace cooperate with US Government (USG) elements?
- What is the approximate wage and economic status of the average citizen?
- What formal and informal educational practices were observed?
  - o Internal threat interference.
  - o Internal threat sponsorship.
- What is the state of health and well-being of the people in this area?
- Did the populace in this area speak the national language differently from others in the country? If so, how?
- What percentage of the populace and the indigenous forces speak English or other foreign languages?
- Did any member of the populace approach or ask questions about US presence or the mission? If so, describe in detail. Give names, if possible.

POLITICAL PARTIES (Major, Minor, or Illegal Parties)

- Targeted by HN or US.
  - For Internal Defense and Development (IDAD) support or as FID target.
  - Lethal, nonlethal, or both.
- Fundamental ideology.
  - Authoritarian or elitist. Populist or democratic.
  - Secular, theocratic, or mixed.
  - Attitude toward HN government.
  - Attitude toward USG.
- Leaders.
  - Key communicators.
  - Willing or unwilling to support psychological operations (PSYOP) program.
  - Effectiveness as spoiler or antigovernment/anti-US firebrand.
- Policies.
- Influence on government.
- Influence on the people.
  - Peaceful/cooperative or militant/front group.
  - Cooperative with PSYOP program.
  - Used in PSYOP series targeting another TA(s).
- Overall effectiveness.
  - Percentage of electorate: claimed vs. actual turnout.
  - Money, real influence, covert influence, spoiler, etc.
- Foreign influence.
  - Ethnic and/or ideological.
  - Regional or international.
  - Stability, strength, and weaknesses.

Figure VI-8. Post Mission Debriefing Guide (continued)

MILITARY
• Friendly forces.
• Disposition.
• Composition, identification, and strength.
• Organization, armament, and equipment.
• Degree of training and combat effectiveness.
• Morale: general and specific:
▪ General psychological strengths and weaknesses.
▪ Degree of stratification—number of TAs.
▪ Psychological vulnerabilities/susceptibilities.
▪ Targeted by HN and/or US PSYOP—effectiveness.
▪ Targeted by internal threat/foreign propaganda—effectiveness.
• Mission.
• Leadership and capabilities of officers and noncommissioned officers compared with those of the United States.
• Logistics.
• Maintenance problems with weapons and equipment.
• Methods of resupply and their effectiveness.
• General relationship between HN military forces, the populace, and other forces (paramilitary, police, etc.).
• Influence on local populace.
▪ Credibility.
▪ Lingering effects of past bad acts/incompetence.
▪ Anecdotal or empirical evidence of improvement(s).
▪ Leaders or rank and file as used as key communicators/disseminators.
▪ Significant operations and/or PSYOP actions with outcomes.
• Recommendation for these forces (military and/or paramilitary) for unconventional warfare contact.
INSURGENT OR OTHER INTERNAL DEFENSE THREAT FORCES*
• Disposition.
• Composition, identification, and strength.
• Organization, armament, and equipment.
• Degree of training, morale, and combat effectiveness.
• Mission.
• Leadership capabilities.
• Logistics.
• Maintenance problems with weapons and equipment.
• Method of resupply and its effectiveness.
• Psychological strengths and weaknesses.
• Relationship between insurgent forces, joint force units, and the populace.
• Influence on local populace.
POLICE AND SECURITY FORCES (Friendly and Adversary)*
• Disposition, strengths, and location.
• Organization, armament, and equipment.
• Logistics.
• Motivation, reliability, and degree of training.
• Psychological strengths and weaknesses.
• Relationship with the government and local populace.

Figure VI-8. Post Mission Debriefing Guide (continued)

AUXILIARY AND UNDERGROUND (Friendly and Adversary) *
• Disposition, strength, and degree of organization.
• Morale and general effectiveness.
• Motivation and reliability.
• Support.
▪ Logistics.
▪ Intelligence.
* Combined in example for brevity. Should be covered separately.

INFRASTRUCTURE Describe the area:
• Rail system.
▪ General route.
▪ Importance to the local and general area.
▪ Bridges, tunnels, curves, and steep grades.
▪ Bypass possibilities.
▪ Key junctions, switching points, and power sources.
▪ Location of maintenance crews who keep the system operational during periods of large-scale interdiction.
▪ Security.
• Telecommunications system.
▪ Location and description of routes, lines, and cables.
▪ Location of power sources.
▪ Location and capacity of switchboards.
▪ Critical points.
▪ Importance to the local general area.
▪ Capabilities of maintenance crews to keep the system operating at a minimum.
▪ Security.
• Petroleum, oils, and lubricants (POL) storage and processing facilities.
▪ Location.
▪ Capacity of storage facilities.
▪ Equipment used for the production of POL.
▪ Power source.
▪ Types and quantities of POL manufactured.
▪ Methods of transportation and distribution.
o Rail.
o Truck.
o Ship.
o Air.
▪ Pipeline routes and pumping station capacities.
▪ Security.
• Electrical power system.
▪ Location and description of power stations.
▪ Principal power lines and transformers.
▪ Location of maintenance crews, facilities, and reaction time.
▪ Critical points.
▪ Capacity (kilowatts).
▪ Principal users.
▪ Security.
• Military installations and depots.
▪ Size.
▪ Activity.
▪ Location.
▪ Units.
▪ Equipment.

Figure VI-8. Post Mission Debriefing Guide (continued)

TARGETS
Describe the area:
• Reaction time.
• Security.
• Highway and road system.
• Name and number.
• Type of surface, width, and condition.
• Location of bridges, tunnels, curves, and steep grades.
• Bypass possibilities.
• Traffic density.
• Location of maintenance crews, facilities, and reaction time.
• Security.
• Inland waterways and canals.
• Name and number.
• Width, depth, and type of bed.
• Direction and speed of flow.
• Location of dams and locks, their power source, and other traffic obstructions.
• Location and descriptions of administrative, control, maintenance crew, facilities, and reaction crew.
• Location and description of navigational aids.
• Natural and synthetic gas system.
• Location and capacity of wells and pipelines.
• Storage facilities and capacity.
• Critical points.
• Maintenance crews, facilities, and reaction time.
• Principal users.
• Security.
• Industrial facilities.
• Capabilities of plants to convert their facilities in wartime to the production of essential military materials.
• Type of facilities.
• Power sources.
• Locations.
• Sources of raw materials.
• Number of employees.
• Disposition of products.
• General working conditions.
• Critical points.
• Security.
• PSYOP actions.
• Restricted targets for cultural, infrastructural, or psychological value.
• Targets requiring nonlethal action.
HEALTH AND SANITATION
• To what degree does hunting and fishing contribute to the local diet?
• What cash crops are raised in the area?
• What domestic and wild animals are present?
• What animal diseases are present?
• What is the availability and quality of water in populated and unpopulated areas?
• What systems are used for sewage disposal?
• What sanitation practices were observed in the populated and unpopulated areas?
• What are the most common human illnesses and how are they controlled?
• What basic health services are available in populated and unpopulated areas?

Figure VI-8. Post Mission Debriefing Guide (continued)

EVASION AND RECOVERY
• From which element of the populace is assistance most likely?
• What, if any, safe houses or areas for evasion and resistance purposes can be recommended?
• What type shelters were used?
• Were fires small and smokeless?
• Were shelters adequate?
• Was food properly prepared?
• Were campsites well chosen?
• Were campsites and trails sterilized after movement to a new one?
• HN/US PSYOP support to evasion and resistance?
CIVIL AFFAIRS OPERATIONS (CAO)
• Has the end state been achieved for CAO supporting civil-military operations (CMO)?
▪ HN transition plan.
▪ Has coordination for handoff been conducted with appropriate commands, agencies, and other organizations?
▪ If no, remaining benchmarks.
• Have the underlying causes of the conflict been ameliorated?
▪ To what degree?
▪ If still existing, how do they influence future planning?
• What arrangements have been made with other organizations to accomplish remaining civil affairs (CA) activities?
• New humanitarian, governmental, and infrastructure assistance requirements during current deployment.
• Will any ongoing operations (for example, engineer projects) be discontinued or interrupted?
• CA functional specialists that remain behind and residual requirements for each:
▪ Rule of law.
▪ Economic stability.
▪ Infrastructure.
▪ Governance.
▪ Public education and information.
▪ Public health and welfare.
• Who will support CA forces that remain behind?
MISCELLANEOUS
• Weather.
▪ Wind speed and direction.
▪ Temperature.
▪ Effect on personnel, equipment, and operations.

Figure VI-8. Post Mission Debriefing Guide (continued)

APPENDIX A
LEGAL CONSIDERATIONS

1. Overview

Unless otherwise stated, the conditions stated in this appendix apply to FID and SFA. Law and policy govern the actions of the US forces in all military operations, including SFA and FID. For US forces to conduct operations, a legal basis must exist. This legal basis profoundly influences many aspects of the operation. It affects the ROE, how US forces organize and train foreign forces, the authority to spend funds to benefit the HN, and the authority of US forces to detain and interrogate. The President is Commander in Chief of the US forces. Therefore, orders issued by the President or SecDef to a CCDR provide the starting point in determining the legal basis. Laws are legislation passed by Congress and signed into law by the President, as well as treaties to which the US is party. Policies are executive orders, departmental directives and regulations, and other authoritative statements issued by government officials. No summary provided here can replace a consultation with the unit's supporting staff judge advocate (SJA). This appendix summarizes some of the laws and policies that bear upon US military operations in support of SFA and FID.

2. Legal Authority for Security Force Assistance and Foreign Internal Defense

Without a deployment or execution order from the President or SecDef, US forces may be authorized to make only limited contributions during operations that involve FID. If the Secretary of State requests and SecDef approves, US forces can participate in FID. The request and approval may go through standing statutory authorities in Title 22, USC. Among other programs, Title 22, USC, contains the FAA and the AECA. Programs under Title 22, USC, authorize SA, developmental assistance, and other forms of aid. The request and approval might also occur under various provisions in Title 10, USC. Title 10, USC, authorizes certain types of military-to-military contacts, exchanges, exercises, and limited forms of HCA in coordination with the US ambassador to the HN. This cooperation and assistance is limited to liaison, contacts, training, equipping, and providing defense articles and services. It does not include direct involvement in operations. Assistance to police by US forces is permitted but not with DOD as the lead government department.

a. Distinguishing Sentiment, Policy, and Law. The underlying international sentiment as to what is acceptable behavior in conflict and war often rapidly outpaces formal treaty adoption and ratification. In short, the collective, largely unwritten will of a majority of the international community can become customary international law. In addition, the USG often formulates and champions this emerging law and policy. In the fight against terrorism, contemporary operational environment, international policy, and sentiment have been greatly debated among the legitimate nation-states of the world, and terrorist forces and sympathizers have attempted to shape and exploit that debate to their advantage. Joint forces conducting FID operations face similar conditions unless and until international law codifies terrorism, insurgency, and other forms of violent lawlessness. Even as such codes, laws, and conventions emerge, joint forces conducting FID operations will likely always face disinformation and propaganda that vilifies legitimate military, reconstruction, and law enforcement efforts as violations of what the adversary will refer to as *international law.*

b. Planning Concerns

(1) Those planning and conducting FID operations may often need a detailed knowledge of international law, such as the Geneva Conventions, for two principal reasons. The first is to educate HN military staffs and forces. The second is to counter very specific points of adversary disinformation and propaganda. Advisors and trainers may have to build either a knowledge base on international law in HN military personnel or an adherence to portions that the HN military has routinely ignored in the past. In addition, this may carry over to transgressions of their own HN laws or building acceptance of new HN laws safeguarding civil liberties. Basic human rights also include other rights, such as the right of free speech, freedom of worship, and freedom of the press, that HN soldiers must uphold in FID operations, because of the concerns of international and US law. US personnel who notice suspected violations of basic human rights must report the facts to their chain of command. Under US law, the President must cut off SA to any country with a documented pattern of human rights abuses.

(2) Internal threat propagandists increasingly use factual, partially factual, or entirely fictitious violations of international law, policy, or even sentiment to discredit HN governments. These attempts are frequently graphic to have the maximum incendiary effect. They often address third countries or international agencies and may cite specific articles of the Geneva Conventions. Citing specific portions of the Geneva Conventions accomplishes two goals for them. If successful, they appear to have legitimate status as a state actor, and they make the HN look like a nation that ignores civil rights and the laws of war. FID forces must infuse an acceptance of the basic tenents of international law among the HN personnel they work with, advise, and train.

3. **Existing United States Law**

a. The United States Constitution. The Constitution divides the power to wage war between the Executive and Legislative branches of government. Under Article I, Congress holds the power to declare war, to raise and support armies, to provide and maintain a navy, and to make all laws necessary and proper for carrying out those responsibilities. Balancing that legislative empowerment, Article II vests the Executive power in the President and makes him the Commander in Chief of the Armed Forces. This bifurcation of the war powers created an area in which the different political branches of government exercise concurrent authority over decisions relating to the use of Armed Forces overseas as an instrument of US foreign policy.

b. The Supremacy Clause of the Constitution (Article VI) states, in part, that all treaties made by the United States are the "supreme law of the land." Therefore, ratified treaties, such as the UN Charter and the Geneva Conventions, create legal obligations on US forces regarding their ability to perform various types of missions and functions.

    c. The War Powers Resolution (WPR) of 1973. The stated purpose of the WPR is to ensure the "collective judgment" of both the Executive and Legislative branches in order to commit to the deployment of US forces by requiring consultation of and reports to Congress, in any of the following circumstances:

        (1) Introduction of troops into actual hostilities.

        (2) Introduction of troops, equipped for combat, into a foreign country, or greatly enlarging the number of troops, equipped for combat, in a foreign country.

Note. The President is required to make such reports within 48 hours of the triggering event, detailing the circumstances necessitating introduction or enlargement of troops, the Constitutional or legislative authority upon which the action is based, and the estimated scope and duration of the deployment or combat action. Since the WPR was passed over the veto of President Nixon and became law, no President has either conceded the constitutionality of the WPR or complied fully with its mandates.

4. International Law and Treaties

    a. The UN Charter became effective on 24 October 1945 after being ratified by the US and a majority of other signatories. The UN Charter mandates that all member states resolve their international disputes peacefully and requires that they refrain in their international relations from the threat or use of force. The UN Charter also provides that all nations have the right to use self-defense to combat acts of aggression against them until such time as the Security Council shall take action.

    b. The UN Charter provides the essential framework of authority for the use of force, effectively defining the foundations for a modern *jus ad bellum* (the law governing a state's resort to force). Inherent in its principles are the requirements for *necessity* (which involves considering the exhaustion or ineffectiveness of peaceful means of resolution, the nature of coercion applied by the aggressor state, objectives of each party, and the likelihood of effective community intervention); *proportionality* (i.e., limiting force in magnitude, scope, and duration to that which is reasonably necessary to counter a threat or attack); and an element of *timeliness* (i.e., delay of a response to an attack or the threat of attack attenuates the immediacy of the threat and the necessity to use force in self-defense).

    c. US forces obey the law of war during all armed conflicts, however such conflicts are characterized, and in all other military operations. The law of war is a body of international treaties and customs, recognized by the US as binding. It regulates the conduct of hostilities and protects noncombatants and civilians. The main laws of war protections come from the Hague and Geneva Conventions.

d. During SFA or FID operations, commanders must be aware of Common Article 3 of the Geneva Conventions and the status of insurgents under the laws of the HN. Common Article 3 is contained in all four of the Geneva Conventions and is specifically intended to apply to internal armed conflicts. Common Article 3 states the following: In the case of armed conflict not of an international character occurring in the territory of one of the high contracting parties, each party to the conflict shall be bound to apply, as a minimum, the following provisions:

(1) Persons taking no active part in the hostilities, including members of armed forces who have laid down their arms and those placed "hors de combat," taken out of the fight, by sickness, wounds, detention, or any other cause, shall in all circumstances be treated humanely, without any adverse distinction founded on race, color, religion or faith, sex, birth or wealth, or any other similar criteria. To this end, the following acts are and shall remain prohibited at any time and in any place whatsoever with respect to the above-mentioned persons:

(a) Violence to life and person, in particular murder of all kinds, mutilation, cruel treatment, and torture;

(b) Taking of hostages;

(c) Outrages upon personal dignity, in particular humiliating and degrading treatment; and

(d) The passing of sentences and the carrying out of executions without previous judgment pronounced by a regularly constituted court, affording all the judicial guarantees which are recognized as indispensable by civilized peoples.

(2) The wounded and sick shall be collected and cared for. An impartial humanitarian body, such as the International Committee of the Red Cross, may offer its services to the parties to the conflict. The parties to the conflict should further endeavor to bring into force, by means of special agreements, all or part of the other provisions of the Geneva Conventions. The application of the preceding provisions shall not affect the legal status of the parties to the conflict.

5. Application of Criminal Laws of the Host Nation

The final sentence of Common Article 3 makes clear that insurgents have no special status under international law. They are not, when captured, prisoners of war. Insurgents may be prosecuted as criminals for bearing arms against the government and for other offenses, so long as they are accorded the minimum protections described in Common Article 3. US forces conducting SFA should remember that the insurgents are, as a legal matter, criminal suspects within the legal system of the HN. Counterinsurgents must carefully preserve weapons, witness statements, photographs, and other evidence collected at the scene. This evidence will be used to process the insurgents into the legal system and thus hold them accountable for their crimes while still promoting the rule of law.

6. Host Nation Law and Status-of-Forces Agreements

a. The military leader must be aware of and consider the impact of other bodies of law that impact the planning and execution phases, including HN law and any applicable SOFAs.

b. SJAs and planners must be familiar with any SOFAs or other similar agreements that may be applicable. In any given mission, there may be agreements short of SOFAs, such as diplomatic notes, on point. Relevant international documents affecting military operations may be difficult to locate. Several sources are available in which to locate applicable international agreements governing the status of US forces or affecting military operations. DOS publications, such as Treaties in Force, contain unclassified international agreements. Both the relevant combatant command's legal office and the DATT or military assistance group at the embassy should also have access to host national or international agreements impacting the military operation.

c. SOFAs and other international agreements establish the legal status of military personnel in foreign countries. Topics that are usually covered in a SOFA include criminal and civil jurisdiction, taxation, and claims for damages and injuries. In the absence of an agreement or some other arrangement with the HN, DOD personnel in foreign countries may be subject to HN laws. It is essential that all personnel understand status of US forces in the area of operations and are trained accordingly.

7. Legal Constraints on the Security Force Assistance and Foreign Internal Defense Mission

US law, regulations, and policy play a key role in establishing the parameters by which military forces may conduct SFA and FID missions. These factors tend to constitute constraints on the activities of military units. They range from the ROE in combat situations to the ability to spend government funds for a training or support mission.

8. General Prohibition on Assistance to Police

Usually, DOD is not the lead government department for assisting foreign governments. DOS is the lead when US forces provide SA—military training, equipment, and defense articles and services—to HN military forces. The FAA specifically prohibits assistance to foreign police forces except within specific exceptions and under a Presidential directive. When providing assistance to training, the DOS's INL provides the lead role in police assistance. The President, however, may delegate this role to other agencies.

9. Training and Equipping Foreign Forces

All training and equipping of FSF must be specifically authorized. US laws require Congress to authorize expenditures for training and equipping foreign forces. US law also requires DOS to verify that the HN receiving the assistance does not commit gross violations of human rights. Usually, DOD involvement is limited to a precise level of man hours and materiel requested by DOS under the FAA. The President may authorize deployed US forces to train or advise HN security forces as part of the mission. In this case, DOD personnel, operations, and maintenance appropriations provide an incidental benefit to those security forces. All other weapons, training, equipment, logistic support, supplies, and services provided to foreign forces by DOD are paid for with funds appropriated by Congress for that purpose. Moreover, the President gives specific authority to DOD for its role in such "train and equip" efforts. Absent such a directive, DOD lacks authority to take the lead in assisting an HN in training and equipping its security forces.

10. Rules of Engagement

ROE are directives issued by competent military authority that delineate the circumstances and limitations under which US forces will initiate and/or continue combat engagement with opposing forces. Often these directives are specific to the operation. If there are no operation-specific ROE, US forces apply SROE. When working with a multinational force, commanders must coordinate the ROE thoroughly.

11. Fiscal Law Considerations

a. In SFA or FID missions, like all operations, commanders require specific authority to expend funds. This authority is normally found in the DOD Appropriations Act. As a general rule, O&M funds may not be used for SFA or FID missions. Congress may appropriate additional funds to commanders for the specific purpose of conducting more complex stability operations that are not typically covered by O&M. Examples include the commander's emergency response program, the Iraq Relief and Reconstruction Fund, Iraq Freedom Fund, and Commander's Humanitarian Relief and Reconstruction.

b. The legal authority for DOS to conduct foreign assistance is found in the FAA, Title 22, USC, Section 2151.

c. There are two exceptions to the general rule requiring the use of Title 22, USC, funds for foreign assistance:

(1) Interoperability, Safety, and Familiarization Training. DOD may fund the training (as opposed to goods and services) of foreign militaries with O&M dollars only when the purpose of the training is to enhance the interoperability, familiarization, and safety training. O&M funds may not be used for SA training. This exception applies only to interoperability training.

OPERATION ENDURING FREEDOM-PHILIPPINES

In January 2002, Joint Task Force (JTF) 510 made up in large part by special operations forces personnel deployed to the southern Philippines to assist the Republic of the Philippines (RP) government in the destruction of the terrorist Abu Sayyaf Group. The initiation of Operation ENDURING FREEDOM-Philippines was only accomplished after considerable debate within the RP government and in the national press. The issue was debated on a constitutional level as the RP constitution banned foreign soldiers on Philippine soil except in time of war. Despite the ardent support of President Corazon Aquino and much of the government and with widespread public support, there was still considerable opposition to the presence of US troops. An eventual compromise had operational effects on JTF 510. A cap was placed on the number of US troops who could deploy, and troops were prohibited from conducting anything other than self-defense. Supplemental understandings to existing agreements were necessary.

Early in the operation, political sensitivities among the populace were challenged when US personnel using an automated teller machine outside their base in Zamboanga City, Mindanao, in uniform and carrying weapons were photographed by the Philippine media. Further restrictions on US personnel were necessary to placate those politically opposed to the US presence. A culturally attuned approach allowed JTF 510 to assist and advise the Armed Forces of the Philippines in early tactical and operational successes against the Abu Sayyaf Group. Slowly, an expansion of the role of US forces, to include combined civil affairs operations, was possible because of early sensitivity to the host nation's laws and political realities.

Various Sources

(2) Congressional Appropriation and Authorization to Conduct Foreign Assistance. DOD may fund foreign assistance operations if Congress has provided a specific appropriation and authorization to execute the mission.

d. The "Leahy Amendment" contains additional constraints on government funding of SFA/FID missions. The law, first enacted in the 1997 Foreign Operations Appropriations Act (the annual DOS appropriations act), prohibits the USG from providing funds to the security forces of a foreign country if DOS has credible evidence that the foreign country or its agents have committed gross violations of human rights, unless the Secretary of State determines and reports that the government of such country is taking effective measures to bring the responsible members of the security forces to justice.

e. Congress specifically appropriates funds for foreign assistance. USAID expends such funds under the legal authorities in Title 22, USC. In addition, provisions of Title 10, USC, authorize small amounts of funds to be appropriated annually for commanders to provide humanitarian relief, disaster relief, or civic assistance in conjunction with military operations. These standing authorities are narrowly defined and generally require significant advance coordination within DOD and DOS.

f. The Coast Guard is specifically authorized to assist other federal agencies in the performance of any activity for which the Coast Guard is especially qualified. As a Service, the Coast Guard has very limited organic legislative authority to unilaterally provide training and technical assistance to foreign countries. With few exceptions, the Coast Guard is generally a service provider to other USG funding agencies whose international engagement authorities convey to the Coast Guard with the transfer of funding, for the specific mission. In accordance with the Economy Act, Title 31, USC, Section 1535, the costs incurred by the Coast Guard while delivering international training and technical assistance are reimbursable whenever the training/assistance is funded by or through another USG or foreign government agency.

KEY SECURITY ASSISTANCE AND FOREIGN ASSISTANCE FUNDING PROGRAMS

The following programs have funds appropriated by Congress to allow the Department of State (DOS) to conduct its foreign assistance mission:

- Foreign Military Financing Program
- International Military Education and Training Program
- Economic Support Fund
- Peacekeeping Operations
- Antiterrorism Assistance
- Global Humanitarian Demining
- Refugee Assistance
- Personnel Details

The following programs are administered by DOS, but do not have funds appropriated to sustain them:

- Foreign Military Sales Program
- Foreign Military Lease Program
- Economy Act Security Assistance
- United States Government Commodities and Services Program
- Direct Commercial Sales Program

There are additional special programs created by law to assist in the foreign assistance mission. These programs include:

- Excess Defense Articles
- Presidential Drawdowns

DOS directly, or indirectly through the United States Agency for International Development, finances numerous development assistance programs to address the following needs:

- Agriculture and Nutrition
- Population Control
- Health
- Education
- Energy
- Environment Improvement

Commander's Emergency Response Program (CERP). CERP is an example of a targeted humanitarian assistance fund program. CERP's primary purpose is "[to enable] military commanders in Iraq [and Afghanistan] to respond to urgent humanitarian relief and reconstruction requirements within their area of responsibility by carrying out programs that will immediately assist the Iraqi [and Afghan] people." CERP was originally funded with seized Iraqi assets, but Congress later appropriated US funds for the purpose. CERP is a program established to assist in missions in Iraq and Afghanistan. It is not applicable to missions outside of those countries. Future missions, though, may have similar funding sources established to facilitate a humanitarian assistance mission. Consult with the servicing judge advocate to determine the availability of funding.

Intentionally Blank

APPENDIX B
JOINT INTELLIGENCE PREPARATION OF THE OPERATIONAL ENVIRONMENT TO SUPPORT FOREIGN INTERNAL DEFENSE

1. Introduction

a. JIPOE is the analytical process used by joint intelligence organizations to produce intelligence assessments, estimates, and other intelligence products in support of the JFC's decision-making process.

b. JIPOE is a four-step process that defines the total operational environment, describes the impact of the operational environment, evaluates the adversary, and determines and describes adversary potential COAs.

For further information on the four-step process for JIPOE, refer to JP 2-01.3, Joint Intelligence Preparation of the Operational Environment.

c. JIPOE for FID, like any operation, is accomplished through a mix of analysis utilizing all the intelligence disciplines. JIPOE utilizes seven intelligence disciplines, namely, GEOINT, HUMINT, signals intelligence, measurement and signature intelligence, open-source intelligence (OSINT), technical intelligence, and CI. The primary purpose of JIPOE is to support joint operation planning, execution, and assessment by identifying, analyzing, and assessing the adversary's centers of gravity, critical vulnerabilities, capabilities, decisive points, limitations, intentions, COAs, and reactions to friendly operations based on a holistic view of the operational environment. The goal is to gain a solid understanding of internal factors as well as regional factors. Accordingly, JIPOE for FID is divided into five categories: OAE, geographic analysis, population analysis, climatology analysis, and threat evaluation.

2. Operational Area Evaluation

a. The JIPOE to support FID begins with a broad OAE, which covers the JFC's operational area. During this phase, data is collected to satisfy basic IRs in the following areas: political, military, economic, religious, social, endemic diseases and health status of the population, geographic, psychological, cultural, friendly forces, threat forces, and nonbelligerent third party forces. Data is collected with respect to the specific operational area and mission and considers all instruments of national power from a strategic perspective. Of particular interest during this stage is the evaluation of the PSYOP and CMO estimate.

b. The PSYOP OAE is initially composed of special PSYOP studies and special PSYOP assessments. These studies identify psychological vulnerabilities, characteristics, insights, and opportunities that exist in the operational area. Analysts doing PSYOP OAE also focus on, but do not limit themselves to, identifying:

(1) The ethnic, racial, social, economic, religious, and linguistic groups of the area and their locations and densities.

(2) Key leaders and communicators in the area, both formal (such as politicians and government officials) and informal (such as businessmen, clergy, or tribal leaders).

(3) Cohesive and divisive issues within a community; as examples, what makes it a community, what would split the community, and what are the attitudes toward the HN.

(4) Literacy rates and levels of education.

(5) Types of media consumed by the community and the level of credibility each is perceived to carry within the community and/or segments within society.

(6) Any concentrations of third country nationals in the operational area and their purposes and functions.

(7) Scientific and technical developments, production, and trade, including significant trade agreements, restrictions, and sanctions, or lack thereof.

(8) The use of natural resources, industry, and agriculture, and destruction or exploitation of the environment.

(9) Location, type, and quantity of toxic industrial material.

c. In the course of OAE, the PSYOP planners coordinate with the military PA office, the cultural officer, and PA staff within the country team to prepare a matrix identifying groups, their leaders, preferred media, and key issues that should be developed. Target groups are identified. The locations of mass media facilities in the area that can be used for the dissemination of PSYOP products, and the identification of their operational characteristics, are also important in the selection of the proper outlet for these products. In particular, the PSYOP planner must evaluate:

(1) Studios and transmitters for radio and television and their operational characteristics (wattage, frequency, and programming).

(2) Heavy and light printing facilities, including locations, types, and capacities of equipment that can supplement the capabilities of PSYOP units.

(3) Accessibility of such facilities to PSYOP forces; as examples, who controls them and whether they will cooperate with the United States.

d. CMO OAE in FID is composed of an evaluation of HN civic action programs, populace and resource control, civilian labor, and materiel procurement. Operational concerns may expand the evaluation to a CMO estimate. The CMO planner also evaluates future sites and programs for civic action undertaken in the operational area by the HN unilaterally or with US support through CMO. In making this evaluation, the planner often relies primarily on the local and regional assets of the HN and the supported command to get an accurate feel for the lawlessness, subversion, insurgency, or other related FID threats that may exist in the area.

3. Geographic Analysis

a. The next JIPOE category to support FID is the geographic analysis, which considers a wide range of factors that include the political, military, economic, religious, social, psychological, and cultural significance of the area. Although not relying exclusively on the two disciplines of GEOINT and OSINT, the geographic analysis accomplished during the JIPOE for a FID operation relies heavily on these two disciplines. The three components of GEOINT, namely imagery, imagery intelligence, and geospatial information, provide the whole of data (or nearly so) on some of the six overlays for FID geographic analysis, such as the cover and concealment overlay, and portions of data for other overlays, such as the logistics sustainment overlay. Similarly, the incident overlay may rely heavily on OSINT data.

b. Normally, the six map overlays described below are a result of the geographic analysis.

(1) **Population Status Overlay.** The generic population status overlay graphically represents the sectors of the population that are pro-government, anti-government, pro-threat, anti-threat, and uncommitted or neutral. This overlay is important because the population can provide support and security to friendly or threat forces. This graphic may also display educational, religious, ethnic, or economic aspects of the population. A more refined product in an urban environment displays the home and work places of key friendly and threat military or civilian personnel and their relatives. In this instance, large-scale maps and/or imagery are used to conveniently plot information by marking rooftops of buildings, although care must be taken to derive coordinates only from non-displaced ground footprints of buildings. Such a refined product should be cross-referenced to order of battle (ORBAT) files that are analogous to the represented data such as personality files and/or faction and organization files.

(2) **Cover and Concealment Overlay.** The cover and concealment overlay graphically depicts the availability, density, type, and location of cover and concealment from the ground as well as from the air. In areas of significant threat of aerial attack or observation, overhead cover and concealment may be important considerations for threat selection of base camps, mission support sites, drug laboratories, or other adversary areas. Surface configuration primarily determines cover, including natural and man-made features such as mines, bunkers, tunnels, and fighting positions. Vegetation is the primary feature that provides concealment. The canopy closure overlay is critical for the determination of areas that offer concealment from aerial observation, particularly in tropical rain forests, and is incorporated into the cover and concealment overlay for rural and other forested areas. In built-up areas, man-made structures are also assessed for the

cover and concealment they offer. When used with the population status overlay, the cover and concealment overlay can be used to determine dwelling and work places, safe houses, routes of movement, and meeting places.

(3) **Logistics Sustainability Overlay.** Logistics is essential to friendly and threat operations. The detection and location of supply lines and bases are critical to finding and defeating hostile activities. Attention is given to basic food, water, medicine, and materiel supply. In rural areas, the logistics sustainability overlay depicts potable water supplies, farms, orchards, growing seasons, and other relevant items. In built-up areas, this overlay depicts supermarkets, food warehouses, pharmacies, hospitals, clinics, and residences of doctors and other key medical personnel. Key to preparing this overlay is knowledge of threat and friendly forces, their logistic requirements, and the availability and location of materiel and personnel to meet these requirements.

(4) **Target Overlay.** The target overlay graphically portrays the location of possible threat targets within the area. In FID environments, this overlay depicts banks, bridges, electric power installations, bulk petroleum and chemical facilities, military and government facilities, the residences and work places of key friendly personnel, and other specific points most susceptible to attack based on threat capabilities and intentions. Hazard estimates are prepared for those targets with collateral damage potential. For example, the threat to a large airbase may focus on airframes, crew billets, and petroleum, oils, and lubricants storage as opposed to runways, aprons, or the control tower. The target overlay is significant to the friendly commander's defensive planning because it shows where defenses need to be concentrated and, conversely, where defenses can be diffused. It also provides CI personnel with a focus for indicators of threat preparation to attack; for example, to discover an indigenous worker pacing off the distances between perimeter fences and critical nodes. The target overlay is useful in disaster relief operations by identifying likely locations for rioting, pilfering, looting, or areas of potential collateral damage.

(5) **Line of Communications Overlay.** The LOC overlay highlights transportation systems and nodes within the area such as railways, roads, trails, navigable waterways, ports, airfields, drop zones (DZs), and landing zones (LZs). In urban environments, mass public transit routes and schedules, as well as underground sewage, drainage and utility tunnels, ditches and culverts, and large open areas that could be used for DZs and LZs are also shown. Where applicable, this overlay will also show seasonal variations. Care is taken to compare recent imagery and geospatial information to ensure that new LOCs are added to the final product. In many situations, LOC products will be readily available from the HN or other local sources.

(6) **Incident Overlay.** The incident overlay plots security related incidents by type and location. Clusters of similar incidents represent a geographic pattern of activity. These incidents can then be further analyzed for time patterns, proximity to population grouping, LOCs, targets, and areas of cover and concealment. This analysis assists in the day-to-day application of security resources.

c. PSYOP and CMO considerations also impact the geographic analysis as described below.

(1) PSYOP considerations in a geographic analysis focus on how geography affects the population of the area and the dissemination of PSYOP products. This step may include, for example, preparation of a radio line of sight (LOS) overlay for radio and television stations derived from an obstacle overlay depicting elevations and LOS information. PSYOP terrain analysis will, for example, focus on determining the respective ranges and audibility of signals from the most significant broadcast stations identified during OAE and locations for cellular communication towers/stations.

(2) CMO considerations in geographic analysis include the identification of critical government, insurgent, and terrorist threats and other threats to food and water storage facilities, toxic industrial material sites, resupply routes, and base locations. In addition, a primary consideration in FID is how terrain affects the ability of US and HN forces to conduct CMO. For example, extremely rugged or thickly vegetated areas may be unsuited to some CMO projects because of inaccessibility to the necessary manpower and equipment needed to initiate such projects.

4. **Population Analysis**

a. In FID, the local population is the key element to successful operations. The CMO planner in conjunction with assigned, attached, or supporting CA forces, is a critical contributor to this element of JIPOE. Elements of the CMO OAE provide the basis for this analysis, as well as CA area study and assessments previously created or created in support of the mission. During this analysis, the planner identifies, evaluates, and makes overlays and other products as appropriate for the following factors: social organization; economic organization and dynamics; political organization and dynamics; history of the society; nature of the insurgency (if applicable); nature of the government; effects on nonbelligerents; and COAs of the insurgents, the HN government, and nonbelligerents.

b. In evaluating social organization, planners look at:

(1) Density and distribution of population by groups, balance between urban and rural groups, sparsely populated areas, and concentrations of primary racial, linguistic, religious, or cultural groups.

(2) Race, religion, national origin, tribe, economic class, political party and affiliation, ideology, education level, union memberships, management class, occupation, and age of the populace.

(3) Overlaps among classes and splits within them; such as the number and types of religious and racial groups to which union members belong and ideological divisions within a profession.

(4) Composite groups based on their political behavior and the component and composite strengths of each; that is, those who actively or passively support the government or the threat, and those who are neutral.

(5) Active or potential issues motivating the political, economic, social, or military behavior of each group and subgroup.

(6) Population growth or decline, age distribution, and changes in location by groups.

(7) Finally, planners perform a factor analysis to determine which activities and programs accommodate the goals of most of the politically and socially active groups. Then they determine which groups and composite groups support, are inclined to support, or remain neutral toward the government.

c. In evaluating economic organization and performance, planners specifically look at:

(1) The principal economic ideology of the society and local innovations or adaptations in the operational area.

(2) The economic infrastructure such as resource locations, scientific and technical capabilities, electric power production and distribution, transport facilities, and communications networks.

(3) Economic performance such as gross national product, gross domestic product, foreign trade balance, per capita income, inflation rate, and annual growth rate.

(4) Major industries and their sustainability including the depth and soundness of the economic base, maximum peak production levels and duration, and storage capacity.

(5) Performance of productive segments such as public and private ownership patterns, concentration and dispersal, and distribution of wealth in agriculture, manufacturing, forestry, information, professional services, mining, and transportation.

(6) Public health factors that include, but are not limited to, birth and death rates, diet and nutrition, water supply, sanitation, health care availability, endemic diseases, health of farm animals, and availability of veterinary services.

(7) Foreign trade patterns such as domestic and foreign indebtedness (public and private) and resource dependencies.

(8) Availability of education including access by individuals and groups, sufficiency for individual needs; groupings by scientific technical, professional, liberal arts, and crafts training; and surpluses and shortages of skills.

(9) Unemployment, underemployment, and exclusion of groups, as well as horizontal and vertical career mobility.

(10) Taxing authorities, rates, and rate determination.

(11) Economic benefit and distribution, occurrence of poverty, and concentration of wealth.

(12) Population shifts and their causes and effects; as examples, rural to urban, agriculture to manufacturing, and manufacturing to service.

(13) Finally, planners identify economic program values and resources that might generate favorable support, stabilize neutral groups, or neutralize threat groups.

d. In evaluating political organization and dynamics, planners specifically look at:

(1) The formal political structure of the government and the sources of its power; that is, pluralist democracy based on the consensus of the voters or strong man rule supported by the military.

(2) The informal political structure of the government and its comparison with the formal structure; that is, is the government nominally a democracy but in reality a political dictatorship?

(3) Legal and illegal political parties and their programs, strengths, and prospects for success. Also, the prospects for partnerships and coalitions between the parties.

(4) Nonparty political organizations, motivating issues, strengths, and parties or programs they support such as political action groups.

(5) Nonpolitical interest groups and the correlation of their interests with political parties or nonparty organizations such as churches, cultural and professional organizations, and unions.

(6) The mechanism for government succession, the integrity of the process, roles of the populace and those in power, regularity of elections, systematic exclusion of identifiable groups, voting blocs, and patron-client determinants of voting.

(7) Independence or subordination and effectiveness of the judiciary. That is, does the judiciary have the power of legislative and executive review? Does the judiciary support constitutionally guaranteed rights and international concepts of human rights?

(8) Independence or control of the press and other mass media and the alternatives for the dissemination of information and opinion.

(9) Centralization or diffusion of essential decisionmaking and patterns of inclusion, or inclusion of specific individuals or groups in the process.

(10) Administrative competence of the bureaucracy. Are bureaucrats egalitarian in practice or in words only? Can individuals and groups make their voices heard within the bureaucracy?

(11) Finally, planners correlate data concerning political, economic, and social groups and then identify political programs to neutralize opposing groups as well as provide programs favorable to friendly groups.

e. In evaluating the history of the society, planners specifically look at:

(1) The origin of the incumbent government and its leadership. Was it elected? Does it have a long history? Have there been multiple peaceful successions of government?

(2) The history of political violence. Is violence a common means for the resolution of political problems? Is there precedent for revolution, *coup d'etat*, assassination, or terrorism? Does the country have a history of consensus-building? Does the present insurgency have causes and aspirations in common with historic political violence?

(3) Finally, the analysts determine the legitimacy of the government, acceptance of violent and nonviolent remedies to political problems by the populace, the type and level of violence to be used by friendly and threat forces, and the groups or subgroups that will support or oppose the use of violence.

f. In evaluating the nature of the insurgency, planners specifically look at:

(1) Desired end state of the insurgency, clarity of its formulation, openness of its articulation, commonality of point of view among the elements of the insurgency, and differences between this end view and the end view of the government.

(2) Groups and subgroups supporting the general objectives of the insurgency.

(3) Divisions, minority views, and dissension within the insurgency.

(4) Groups that may have been deceived by the threat concerning the desired end state of the insurgency.

(5) Organizational and operational patterns used by the insurgency, variations and combinations of such, and shifts and trends.

(6) Finally, analysts determine the stage and phase of the insurgency as well as how far and how long it has progressed and/or regressed over time. They identify unity and disagreement with front groups, leadership, tactics, primary targets, doctrine, training, morale, discipline, operational capabilities, and materiel resources. They evaluate external support, to include political, financial, and logistic assistance. This should include not only the sources of support, but also specific means by which support is provided and critical points through which the HN could slow, reduce, negate, or stop this support. The planners determine whether rigid commitment to a method or ideological tenet or other factor constitutes an exploitable vulnerability and/or a weakness on which the government can build strength.

g. When examining hostile groups, planners examine from hostile perspectives:

(1) The leadership and staff structure and its psychological characteristics, skills, and C2 resources.

(2) Patterns of lawless activities (as examples, illicit drug trafficking, extortion, piracy, and smuggling) or insurgent operations, base areas, LOCs, and supporters outside of the country concerned.

(3) The intelligence, OPSEC, deception, and PSYOP capabilities of the hostile groups.

(4) The appeal of the hostile groups to those who support them.

h. In evaluating the nature of the government response, planners specifically examine:

(1) General planning, or lack of planning, for countering the insurgency, lawlessness, or subversion being encountered as well as planning comprehensiveness and correctness of definitions and conclusions.

(2) Organization and methods for strategic and operational planning and execution of plans such as resource requirements, constraints, and realistic priorities.

(3) Use of population and resources and the effects on each group.

(4) Organization, equipment, and tactical doctrine for security forces; for example, how the government protects its economic and political infrastructure.

(5) Areas where the government has maintained the initiative.

(6) Population and resource control measures.

(7) Economic development programs.

(8) Finally, planners correlate government and insurgent strengths and weaknesses and identify necessary changes in friendly programs, plans, organization, and doctrine.

i. In evaluating the effects on nonbelligerents, planners specifically examine:

(1) Mechanisms for monitoring nonbelligerent attitudes and responses.

(2) Common objectives of groups neither supporting nor opposing the insurgency.

(3) Effects on the populace of government military, political, economic, and social operations and programs. That is, does the government often kill civilians in its counterthreat operations? Are benefits of government aid programs evenly distributed?

(4) To whom is the populace inclined to provide intelligence?

(5) Finally, planners determine the strengths and weaknesses of the nonbelligerents, the depth of their commitment to remain neutral, and the requirements to make them remain neutral and/or to support friendly or threat programs or forces.

j. In evaluating COAs for threat forces, the government, and nonbelligerents, analysts balance the foregoing factors and determine likely COAs, as well as the probable outcomes for each element.

5. **Climatology Analysis**

Relevant weather factors extend beyond short-term weather analysis to consideration of the broader and longer term climatological factors. The area's climate, weather, and light conditions are analyzed to determine their effects on friendly, threat, and nonbelligerent third-party operations. Planners consider climate types by area and season and their effects on military, political, social, and economic activities. Historic weather data and weather effects overlays are developed during this step. The effects of weather and climate are integrated with terrain analysis. Special considerations are made for the effects of weather and climate on CMO projects, PSYOP media and dissemination, amounts of accessible food, storage of explosives, and population patterns such as seasonal employment. Examples of potential effects are periods of drought that force farmers to become bandits or insurgents, and flooding that causes isolation and interference with the distribution of food and medicine.

6. **Threat Evaluation**

a. In conducting the threat evaluation in FID, particular attention is paid to the HN government's military and paramilitary police forces and the insurgent forces and infrastructure (guerrilla, auxiliary, and underground). Correlation of force evaluation in such environments includes a detailed analysis of the following factors for friendly, threat, and nonbelligerent forces: composition, strength (include number of active members, amount of popular support, funding method, and origin), training, equipment, electronics technical data, disposition (location), tactics and methods, operational effectiveness, weaknesses and vulnerabilities, personalities, and miscellaneous data.

b. The FID planners determine how the friendly, threat, and nonbelligerent forces can use geography, offensive actions, security, surprise, and cross-country mobility to develop locally superior application of one or more of the instruments of national power. FID planners identify the strengths and weaknesses of friendly, threat, and nonbelligerent forces, and determine the political, social, economic, and psychological effects of each side's COAs, tactics, and countertactics. Finally, the planners develop COAs that will optimize the application of the elements of national power by the friendly side.

c. The PSYOP threat evaluation serves two purposes. First, it provides the commander with an understanding of the existing and potential opposing propaganda in the area. It is a safe assumption that if US forces are conducting PSYOP in an area, some other organization is also conducting PSYOP in the area. US PSYOP forces in the area must anticipate and be able to counter, if not prevent, threat propaganda directed at US and allied forces and the local populace. Second, the PSYOP threat evaluation provides the supported commander with the PSYOP consequences of US operations, and also provides alternative measures within each COA. To conduct an effective threat evaluation, the planner must determine the capabilities of threat organizations to conduct propaganda operations and to counteract US and allied PSYOP. (The demographics of any military or paramilitary threat should be evaluated at this step if they were not considered during OAE.) Specific capabilities to be evaluated include threat abilities to:

(1) Conduct offensive propaganda operations targeting US or allied forces or the local populace.

(2) Indoctrinate its personnel against US PSYOP efforts (defensive counterpropaganda).

(3) Counteract US PSYOP efforts by exploiting weaknesses in US PSYOP operations (offensive counterpropaganda).

(4) Conduct active measures.

(5) Conduct electronic attack against US or allied PSYOP broadcasts.

(6) Conduct electronic protection to safeguard organic PSYOP capability.

d. The CMO threat evaluation focuses on determining the adversaries in the HN population. This determination is especially critical when the opponent is not a standing military force or when the opposing force is not equipped with standard uniforms and weapons such as guerrillas or terrorists. These forces often blend into, or intermingle with, the civilian community. CMO threat evaluation identifies the threat, ORBAT, and modus operandi. Social, religious, and other types of fora through which threat forces employ the instruments of national power, as well as methods of countering such applications, are also identified.

APPENDIX C
ILLUSTRATIVE INTERAGENCY PLAN FOR FOREIGN INTERNAL DEFENSE

1. Purpose

Illustrative interagency plans for FID accomplish the following.

a. Employ all instruments of US national power (diplomatic, informational, military, economic) in support of an HN IDAD effort.

b. Identify and sequence a checklist of taskings for each USG agency over time.

c. Provide a mechanism for USG programs to be mutually supporting.

d. Include clear MOEs and measures of performance working toward clearly defined goals.

e. Integrate USG activities with those of HN and other interested parties.

f. Justify future budget requirements.

g. Establish clear criteria for transition of phases.

h. Inform and guide agency future strategies and plans.

i. Provide a final deconfliction of disparate or contradictory actions by individual agencies.

j. Ensure all agencies communicate the same policies and strategic communication themes.

2. Content

Because there is no set format for an illustrative interagency plan, the commander or lead agent for FID operations would set the form to use. The illustrative interagency plan would include the following components:

a. **Policy Planning Guidance.** This section summarizes guidance provided by the President or other national security decisions pertaining to this situation. This section may include the guidance of the US ambassador. Limiting treaties or further policy guidance from HN policy makers or coalition partners (as endorsed by the President or the President's designee) should be included in this section as well.

b. **United States Interests at Stake.** This section states the US interests at stake that warrant US FID assistance. Clearly stated, transparent motives devoid of any unstated or hidden agendas greatly facilitate the application of the informational and diplomatic instruments of national power. Examples include significant economic interests, reducing international criminal activities affecting US interests, and promoting the spread of human rights and democracy.

c. **United States Strategic Purpose.** This section describes the overall purpose of conducting FID operations in the HN. Increasingly, the strategic purpose of FID transcends the borders of the HN to encompass a regional, transregional, or global strategic purpose. Examples include stabilizing a country for the sake of regional stability, countering narcoterrorism, reducing the potential for proliferation of WMD and the materiel, technology, and expertise necessary to create and sustain a WMD program, and stabilizing a country so that it does not become a haven for terrorists.

d. **Mission Statement.** This section states the who, what, where, when, and why of a USG FID operation. Although not necessarily providing an exhaustive list of the *who,* that is the agencies involved; this mission statement should be in sufficient detail to encompass generalities on how the instruments of national power are going to be brought to bear so that each agency involved in FID operations can infer its level of participation. A comprehensive overview is given with significant operations or known hard dates. For instance, the mission statement may include mention that during the FID operation national elections will occur on a specified date. The *why* of the mission statement should be one of the most complete and least general portions of the mission statement and should comply with and encapsulate US policy, interests, and strategic purpose in the context of the HN IDAD needs.

e. **Desired End State.** This section describes the desired outcome of all FID assistance. The preparer should describe the end state in measurable and quantifiable terms rather than generalities. An example might be a situation where an HN is stabilized to the point that an insurgency is reduced from a national security threat to a minor law enforcement problem. In this case, the level of insurgents might be quantified by a specific number (expressed in an acceptable plus or minus range) coupled with an objectively verifiable metric, such as the actual number of insurgent attacks.

f. **Operational Concept.** This section describes in broad terms how the USG will employ the instruments of national power in the FID operation. The operational concept is not the equivalent of a military concept of operation. The level of detail within an illustrative interagency plan operational concept will typically be less than in a concept of operation. It will instead focus on a holistic description of the interaction of the agencies involved in the FID operation to leverage all instruments of national power to effect political-military conditions in the HN.

g. **Phases.** This section describes phases of USG assistance to an HN. Examples might be support to an HN's transition to a new strategy, support to an HN's operations to regain the initiative, support to HN offensive operations, support to HN consolidation of COIN gains, and rehabilitation. Each phase includes triggers or transition points for movement to the next phase. An alternative type of phasing could be geographical; for instance, pacifying the eastern three regions of a country, then the center, then the west.

h. **Lines of Operation and Political-Military Objectives.** A line of operations (LOO) is a logical line that connects actions on nodes and/or decisive points related in time and purpose with an objective or a physical line that defines the interior or exterior orientation of the force in relation to the enemy or that connects actions on nodes and/or decisive points related in time and space to objectives. This section describes the broad categories of FID activities that the USG will conduct and the objectives within each. An example of a LOO is support to HN security forces to enhance their capacity to deal with insurgency. Political-military objectives within that LOO could include training and equipping a counterguerrilla brigade, supplying 100 helicopters to an HN army, conducting intelligence cooperation, and training the HN police to defend their stations against guerrilla attacks.

i. **Agency Responsibilities.** This section outlines the primary responsibilities of each USG agency involved in this FID operation. In the case of multiple agencies having areas of commonality, limiting or delineating points may be established. In addition, this section can include information on which is the lead and coordinating agency for a line of action that involves multiple agencies.

j. **Implementation Matrix.** This section displays the political-military objectives for each phase in matrix form. Although there is no set format for the implementation matrix, all matrices must show objectives in chronological order. This is true whether the matrix displays hard dates or not. In this case, the matrix shows items in the order they will take place.

k. **Lines of Operations Annexes.** Annexes contain key tasks, each with its MOE, measure of performance, costs, and issues for each LOO. Annexes may be broken down into consolidated multiagency annexes by LOO or by individual agency listing all relevant LOOs for that agency.

Intentionally Blank

APPENDIX D
REFERENCES

The development of JP 3-22 is based upon the following primary references.

1. **Public Laws**

    a. Title 10, USC.

    b. Title 22, USC.

2. **Strategic Guidance and Policy**

    a. National Security Directive Decision 38, *Staffing at Diplomatic Missions and Their Constituent Post.*

    b. National Security Presidential Directive (NSPD) 1, *Organization of the National Security Council System.*

    c. NSPD 8, *National Director and Deputy National Security Advisor for Combating Terrorism.*

    d. NSPD 44, *Management of Interagency Efforts Concerning Reconstruction and Stabilization.*

    e. *National Security Strategy of the United States of America.*

    f. *Guidance for Employment of the Force.*

    g. *National Military Strategy of the United States of America.*

    h. *Joint Strategic Capabilities Plan.*

    i. *National Defense Strategy.*

3. **Department of Defense Publications**

    a. DODD 2000.12, *DOD Antiterrorism Protection.*

    b. DODD 2055.3, *Manning of Security Assistance Organizations and the Selection and USDP Training of Security Assistance Personnel.*

    c. DODD 3000.05, *Military Support for Stability, Security, Transition, and Reconstruction Operations.*

    d. DODD 3000.07, *Irregular Warfare.*

e. DODD 5132.03, *DOD Policy and Responsibilities Relating to Security Cooperation.*

f. DODD 5230.11, *Disclosure of Classified Military Information to Foreign Governments and International Organizations.*

g. DODD 5240.2, *Counterintelligence.*

h. DODI 2205.02, *Humanitarian and Civic Assistance (HCA) Activities.*

i. DODI 3020.37, *Continuation of Essential DOD Contractor Services During Crises.*

j. DODI 5105.81, *Implementing Instructions for DOD Operations at US Embassies.*

k. DOD 5105.38-M, *Security Assistance Management Manual.*

l. Office of the Under Secretary of Defense for Policy, *Guidance for Employment of the Force.*

m. Office of the Under Secretary of Defense for Policy, *Guidance for Development of the Force.*

n. *Memorandum of Agreement Between the Department of Defense and the Department of Homeland Security on the Use of the US Coast Guard Capabilities and Resources in Support of the National Military Strategy.*

4. Chairman of the Joint Chiefs of Staff Publications

a. JP 1, *Doctrine for the Armed Forces of the United States.*

b. JP 1-0, *Personnel Support to Joint Operations.*

c. JP 1-02, *Department of Defense Dictionary of Military and Associated Terms.*

d. JP 2-0, *Joint Intelligence.*

e. JP 2-01.2, *Counterintelligence and Human Intelligence Support to Joint Operations.*

f. JP 2-01.3, *Joint Intelligence Preparation of the Operational Environment.*

g. JP 2-03, *Geospatial Intelligence Support to Joint Operations.*

h. JP 3-0, *Joint Operations.*

i. JP 3-05, *Joint Special Operations.*

j. JP 3-05.1, *Joint Special Operations Task Force Operations.*

k. JP 3-07.4, *Joint Counterdrug Operations.*

l. JP 3-08, *Interorganizational Coordination During Joint Operations.*

m. JP 3-13, *Information Operations.*

n. JP 3-13.2, *Psychological Operations.*

o. JP 3-13.3, *Operations Security.*

p. JP 3-24, *Counterinsurgency Operations.*

q. JP 3-26, *Counterterrorism.*

r. JP 3-33, *Joint Task Force Headquarters.*

s. JP 3-34, *Joint Engineer Operations.*

t. JP 3-57, *Civil-Military Operations.*

u. JP 4-0, *Joint Logistics.*

v. JP 4-02, *Health Service Support.*

w. JP 5-0, *Joint Operation Planning.*

x. JP 6-0, *Joint Communications System.*

y. CJCSI 3110.05C, *Joint Psychological Operations Supplement to the Joint Strategic Capabilities Plan.*

z. CJCSI 3110.05C-1, *Joint Psychological Operations (Classified) Supplement to the Joint Strategic Capabilities Plan.*

aa. CJCSI 3110.12D, *Civil Affairs Supplement to the Joint Strategic Capabilities Plan.*

bb. CJCSI 3121.01B, *Standing Rules of Engagement/Standing Rules for the Use of Force for US Forces.*

cc. CJCSM 3122.03C, *Joint Operation Planning and Execution System, Volume II (Planning Formats).*

5. Multi-Service Publications

a. FM 3-07.10 / Marine Corps Reference Publication 3-33.8A / Navy Tactics, Techniques, and Procedures (NTTP) 3-07.5 / Air Force Tactics, Techniques, and Procedures 3-2.76, *Multi-Service Tactics, Techniques, and Procedures for Training Security Force Advisor Teams.*

b. Navy Warfare Publication (NWP) 1-14M, Marine Corps Warfighting Publication (MCWP) 5-12.1, Commandant of the Coast Guard Publication P5800.7A, *The Commander's Handbook on the Law of Naval Operations.*

6. United States Army Publications

a. FM 3-0, *Operations.*

b. FM 3-05, *Army Special Operations Forces.*

c. FM 3-05.40, *Civil Affairs Operations.*

d. FM 3-05.137, *Army Special Operations Forces Foreign Internal Defense.*

e. FM 3-05.140, *Army Special Operations Forces Logistics.*

f. FM 3-05.202, *Special Forces Foreign Internal Defense.*

g. FM 3-05.301, *Psychological Operations Process, Tactics, Techniques, and Procedures.*

h. FM 3-05.302, *Tactical Psychological Operations Tactics, Techniques, and Procedures.*

i. FM 3-05.401, *Civil Affairs Tactics, Techniques, and Procedures.*

j. FM 3-07, *Stability Operations.*

k. FM 3-24, *Counterinsurgency.*

l. FM 3-24.2, *Tactics for Counterinsurgency.*

m. FM 8-42, *Combat Health Support in Stability Operations and Support Operations.*

n. FM 46-1, *Public Affairs Operations.*

7. **United States Navy Publications**

    a. Navy Doctrine Publication (NDP) 1, *Naval Warfare*.

    b. NDP 5, *Naval Planning*.

    c. NWP 3-05, *Naval Special Warfare*.

    d. NWP 3-06.1, *Navy Riverine and Coastal Operations*.

    e. NTTP 3-54.3, *Operations Security*.

8. **United States Marine Corps Publications**

    a. Fleet Marine Force Reference Publication 12-15, *Small Wars Manual*.

    b. MCWP 3-33A, *Counterguerrilla Operations*.

    c. MCWP 3-33.5, *Counterinsurgency*.

    d. Marine Corps Information Publication 3-33.01, *Small Leader's Guide to Counterinsurgency*.

9. **United States Air Force Publications**

    a. AFDD 2-3, *Irregular Warfare*.

    b. AFDD 2-3.1, *Foreign Internal Defense*.

Intentionally Blank

APPENDIX E
ADMINISTRATIVE INSTRUCTIONS

1. User Comments

Users in the field are highly encouraged to submit comments on this publication to: Commander, United States Joint Forces Command, Joint Warfighting Center, ATTN: Doctrine and Education Group, 116 Lake View Parkway, Suffolk, VA 23435-2697. These comments should address content (accuracy, usefulness, consistency, and organization), writing, and appearance.

2. Authorship

The lead agent for this publication is USSOCOM, and the Joint Staff doctrine sponsor for this publication is the Director for Strategic Plans and Policy (J-5).

3. Supersession

This publication supersedes JP 3-07.1, 30 April 2004, *Joint Tactics, Techniques, and Procedures for Foreign Internal Defense (FID)*.

4. Change Recommendations

a. Recommendations for urgent changes to this publication should be submitted:

```
TO:     CDRUSSOCOM MACDILL AFB FL//SOKF-J7-DD//
INFO:   JOINT STAFF WASHINGTON DC//J7-JEDD//
        JOINT STAFF WASHINGTON DC//J4//
        CDRUSJFCOM NORFOLK VA//DOC GP//
```

Routine changes should be submitted electronically to Commander, Joint Warfighting Center, Doctrine and Education Group and info the Lead Agent and the Director for Operational Plans and Joint Force Development J-7/Joint Education and Doctrine Division via the CJCS Joint Electronic Library (JEL) at http://www.dtic.mil/doctrine.

b. When a Joint Staff directorate submits a proposal to the CJCS that would change source document information reflected in this publication, that directorate will include a proposed change to this publication as an enclosure to its proposal. The Military Services and other organizations are requested to notify the Joint Staff/J-7 when changes to source documents reflected in this publication are initiated.

c. Record of Changes:

CHANGE NUMBER	COPY NUMBER	DATE OF CHANGE	DATE ENTERED	POSTED BY	REMARKS

5. Distribution of Publications

Local reproduction is authorized and access to unclassified publications is unrestricted. However, access to and reproduction authorization for classified JPs must be in accordance with DOD 5200.1-R, *Information Security Program.*

6. Distribution of Electronic Publications

a. Joint Staff J-7 will not print copies of JPs for distribution. Electronic versions are available on the Joint Doctrine, Education, and Training Electronic Information System at https://jdeis.js.mil (NIPRNET), and https://jdeis.js.smil.mil (SIPRNET) and on the JEL at http://www.dtic.mil/doctrine (NIPRNET).

b. Only approved JPs and joint test publications are releasable outside the combatant commands, Services, and Joint Staff. Release of any classified JP to foreign governments or foreign nationals must be requested through the local embassy (Defense Attaché Office) to DIA, Defense Foreign Liaison/IE-3, 200 MacDill Blvd., Bolling AFB, Washington, DC 20340-5100.

c. CD-ROM. Upon request of a joint doctrine development community member, the Joint Staff/J-7 will produce and deliver one CD-ROM with current JPs.

GLOSSARY
PART I — ABBREVIATIONS AND ACRONYMS

AC	Active Component
ACC	area coordination center
ACSA	acquisition and cross-servicing agreement
AECA	Arms Export Control Act
AFDD	Air Force doctrine document
AFSOF	Air Force special operations forces
AOR	area of responsibility
ARSOF	Army special operations forces
ASD(SO/LIC&IC)	Assistant Secretary of Defense for Special Operations and Low-Intensity Conflict and Interdependent Capabilities
AT	antiterrorism
C2	command and control
CA	civil affairs
CAO	civil affairs operations
CBRN	chemical, biological, radiological, and nuclear
CCDR	combatant commander
CD	counterdrug
CDRUSSOCOM	Commander, United States Special Operations Command
CF	conventional forces
CI	counterintelligence
CIA	Central Intelligence Agency
CISO	counterintelligence support officer
CJCS	Chairman of the Joint Chiefs of Staff
CJCSI	Chairman of the Joint Chiefs of Staff instruction
CJCSM	Chairman of the Joint Chiefs of Staff manual
CMO	civil-military operations
COA	course of action
COIN	counterinsurgency
COM	chief of mission
CT	counterterrorism
CTEP	combined training and education plan
CWMD	combating weapons of mass destruction
D&M	detection and monitoring
DATT	defense attaché
D/CIA	Director, Central Intelligence Agency
DEA	Drug Enforcement Administration
DNI	Director of National Intelligence
DOD	Department of Defense
DODD	Department of Defense directive
DODI	Department of Defense instruction

DOJ	Department of Justice
DOS	Department of State
DSCA	Defense Security Cooperation Agency
DSPD	defense support to public diplomacy
DZ	drop zone
EDA	excess defense articles
ETSS	extended training service specialist
FAA	Foreign Assistance Act
FAO	foreign area officer
FHA	foreign humanitarian assistance
FID	foreign internal defense
FM	field manual (Army)
FMF	foreign military financing
FMS	foreign military sales
FP	force protection
FPD	force protection detachment
FSF	foreign security forces
GCC	geographic combatant commander
GEF	Guidance for Employment of the Force
GEOINT	geospatial intelligence
GFS	global fleet station
HCA	humanitarian and civic assistance
HN	host nation
HNS	host-nation support
HSS	health service support
HUMINT	human intelligence
IDAD	internal defense and development
IGO	intergovernmental organization
IIP	Bureau of International Information Programs (DOS)
IMET	international military education and training
INL	Bureau of International Narcotics and Law Enforcement Affairs (DOS)
INTERPOL	International Criminal Police Organization
IO	information operations
IR	intelligence requirement
ISR	intelligence, surveillance, and reconnaissance
IW	irregular warfare
J-2	intelligence directorate of a joint staff
J-3	operations directorate of a joint staff
J-4	logistics directorate of a joint staff

J-5	plans directorate of a joint staff
JEL	Joint Electronic Library
JFC	joint force commander
JIACG	joint interagency coordination group
JIPOE	joint intelligence preparation of the operational environment
JOPES	Joint Operation Planning and Execution System
JP	joint publication
JSCP	Joint Strategic Capabilities Plan
JSOTF	joint special operations task force
JSPS	Joint Strategic Planning System
JTF	joint task force
LOC	line of communications
LOO	line of operations
LOS	line of sight
LZ	landing zone
MAGTF	Marine air-ground task force
MARSOC	Marine Corps special operations command
MARSOF	Marine Corps special operations forces
MCA	military civic action
MCAG	maritime civil affairs group
MCMO	medical civil-military operations
MCWP	Marine Corps warfighting publication
MOE	measure of effectiveness
MSOAG	Marine special operations advisor group
MTT	mobile training team
NA	nation assistance
NDS	national defense strategy
NGO	nongovernmental organization
NMS	National Military Strategy
NSC	National Security Council
NSPD	national security Presidential directive
NSS	National Security Strategy
NTTP	Navy tactics, techniques, and procedures
NWP	Navy warfare publication
O&M	operation and maintenance
OAE	operational area evaluation
OFDA	Office of US Foreign Disaster Assistance
OGA	other government agency
ONDCP	Office of National Drug Control Policy
OPCON	operational control
OPLAN	operation plan

OPSEC	operations security
ORBAT	order of battle
OSD	Office of the Secretary of Defense
OSINT	open-source intelligence
PA	public affairs
PAO	public affairs officer
PCS	permanent change of station
PDSS	predeployment site survey
PEP	personnel exchange program
PIR	priority intelligence requirement
PKO	peacekeeping operations
PM	Bureau of Political-Military Affairs (DOS)
PO	peace operations
POLAD	political advisor
PSYOP	psychological operations
QAT	quality assurance team
RC	Reserve Component
RDCTFP	Regional Defense Combating Terrorism Fellowship Program
ROE	rules of engagement
SA	security assistance
SC	security cooperation
SCO	security cooperation organization
SDO	senior defense official
SecDef	Secretary of Defense
SF	special forces
SFA	security force assistance
SJA	staff judge advocate
SO	special operations
SOC	special operations commander
SOF	special operations forces
SOFA	status-of-forces agreement
SROE	standing rules of engagement
TA	target audience
TAFT	technical assistance field team
TAT	technical assistance team
TCP	theater campaign plan
TDY	temporary duty
TREAS	Department of the Treasury
TSOC	theater special operations command

USAID	United States Agency for International Development
USC	United States Code
USCG	United States Coast Guard
USDAO	United States defense attaché office
USD(P)	Under Secretary of Defense for Policy
USG	United States Government
USJFCOM	United States Joint Forces Command
USN	United States Navy
USSOCOM	United States Special Operations Command
USSOUTHCOM	United States Southern Command
WMD	weapons of mass destruction
WPR	War Powers Resolution

PART II — TERMS AND DEFINITIONS

Unless otherwise annotated, this publication is the proponent for all terms and definitions found in the glossary. Upon approval, JP 1-02, *Department of Defense Dictionary of Military and Associated Terms*, will reflect this publication as the source document for these terms and definitions.

antiterrorism. Defensive measures used to reduce the vulnerability of individuals and property to terrorist acts, to include limited response and containment by local military and civilian forces. Also called **AT.** (JP 1-02. SOURCE: JP 3-07.2)

campaign plan. A joint operation plan for a series of related major military operations aimed at achieving strategic or operational objectives within a given time and space. (JP 1-02. SOURCE: JP 5-0)

civil administration. An administration established by a foreign government in (1) friendly territory, under an agreement with the government of the area concerned, to exercise certain authority normally the function of the local government; or (2) hostile territory, occupied by United States forces, where a foreign government exercises executive, legislative, and judicial authority until an indigenous civil government can be established. Also called **CA.** (JP 1-02. SOURCE: JP 3-05)

civil affairs. Designated Active and Reserve Component forces and units organized, trained, and equipped specifically to conduct civil affairs operations and to support civil-military operations. Also called **CA.** (JP 1-02. SOURCE: JP 3-57)

civil affairs operations. Those military operations conducted by civil affairs forces that (1) enhance the relationship between military forces and civil authorities in localities where military forces are present; (2) require coordination with other interagency organizations, intergovernmental organizations, nongovernmental organizations, indigenous populations and institutions, and the private sector; and (3) involve application of functional specialty skills that normally are the responsibility of civil government to enhance the conduct of civil-military operations. Also called **CAO.** (JP 1-02. SOURCE: JP 3-57)

civil-military operations. The activities of a commander that establish, maintain, influence, or exploit relations between military forces, governmental and nongovernmental civilian organizations and authorities, and the civilian populace in a friendly, neutral, or hostile operational area in order to facilitate military operations, to consolidate and achieve operational US objectives. Civil-military operations may include performance by military forces of activities and functions normally the responsibility of the local, regional, or national government. These activities may occur prior to, during, or subsequent to other military actions. They may also occur, if directed, in the absence of other military operations. Civil-military operations may be performed by designated civil affairs, by other military forces, or by a combination of civil affairs and other forces. Also called **CMO.** (JP 1-02. SOURCE: JP 3-57)

conventional forces. 1. Those forces capable of conducting operations using nonnuclear weapons. 2. Those forces other than designated special operations forces. Also called **CF.** (JP 1-02. SOURCE: JP 3-05)

counterinsurgency. Comprehensive civilian and military efforts taken to defeat an insurgency and to address any core grievances. Also called **COIN.** (JP 1-02. SOURCE: JP 3-24)

counterterrorism. Actions taken directly against terrorist networks and indirectly to influence and render global and regional environments inhospitable to terrorist networks. Also called **CT.** (JP 1-02. SOURCE: JP 3-26)

country team. The senior, in-country, US coordinating and supervising body, headed by the chief of the US diplomatic mission, and composed of the senior member of each represented US department or agency, as desired by the chief of the US diplomatic mission. (JP 1-02. SOURCE: JP 3-07.4)

force protection. Preventive measures taken to mitigate hostile actions against Department of Defense personnel (to include family members), resources, facilities, and critical information. Force protection does not include actions to defeat the enemy or protect against accidents, weather, or disease. Also called **FP.** (JP 1-02. SOURCE: JP 3-0)

foreign humanitarian assistance. Department of Defense activities, normally in support of the United States Agency for International Development or Department of State, conducted outside the United States, its territories, and possessions to relieve or reduce human suffering, disease, hunger, or privation. Also called **FHA.** (JP 1-02. SOURCE: JP 3-29)

foreign internal defense. Participation by civilian and military agencies of a government in any of the action programs taken by another government or other designated organization to free and protect its society from subversion, lawlessness, insurgency, terrorism, and other threats to its security. Also called **FID.** (Approved for incorporation into JP 1-02.)

host nation. A nation which receives the forces and/or supplies of allied nations and/or NATO organizations to be located on, to operate in, or to transit through its territory. Also called **HN.** (JP 1-02. SOURCE: JP 3-57)

host-nation support. Civil and/or military assistance rendered by a nation to foreign forces within its territory during peacetime, crises or emergencies, or war based on agreements mutually concluded between nations. Also called **HNS.** (JP 1-02. SOURCE: JP 4-0)

humanitarian and civic assistance. Assistance to the local populace provided by predominantly US forces in conjunction with military operations and exercises. This assistance is specifically authorized by Title 10, United States Code, Section 401, and funded under separate authorities. Also called **HCA.** (JP 1-02. SOURCE: JP 3-29)

humanitarian assistance. Programs conducted to relieve or reduce the results of natural or manmade disasters or other endemic conditions such as human pain, disease, hunger, or privation that might present a serious threat to life or that can result in great damage to or loss of property. Humanitarian assistance provided by US forces is limited in scope and duration. The assistance provided is designed to supplement or complement the efforts of the host nation civil authorities or agencies that may have the primary responsibility for providing humanitarian assistance. Also called **HA.** (JP 1-02. SOURCE: JP 3-57)

information operations. The integrated employment of the core capabilities of electronic warfare, computer network operations, psychological operations, military deception, and operations security, in concert with specified supporting and related capabilities, to influence, disrupt, corrupt or usurp adversarial human and automated decision making while protecting our own. Also called **IO.** (JP 1-02. SOURCE: JP 3-13)

insurgency. The organized use of subversion and violence by a group or movement that seeks to overthrow or force change of a governing authority. Insurgency can also refer to the group itself. (JP 1-02. SOURCE: JP 3-24)

interagency. United States Government agencies and departments, including the Department of Defense. (JP 1-02. SOURCE: JP 3-08)

interagency coordination. Within the context of Department of Defense involvement, the coordination that occurs between elements of Department of Defense, and engaged US Government agencies for the purpose of achieving an objective. (JP 1-02. SOURCE: JP 3-0)

intergovernmental organization. An organization created by a formal agreement (e.g., a treaty) between two or more governments. It may be established on a global, regional, or functional basis for wide-ranging or narrowly defined purposes. Formed to protect and promote national interests shared by member states. Examples include the United Nations, North Atlantic Treaty Organization, and the African Union. Also called **IGO.** (JP 1-02. SOURCE: JP 3-08)

internal defense and development. The full range of measures taken by a nation to promote its growth and to protect itself from subversion, lawlessness, insurgency, terrorism, and other threats to its security. Also called **IDAD.** (Approved for incorporation into JP 1-02.)

joint intelligence preparation of the operational environment. The analytical process used by joint intelligence organizations to produce intelligence estimates and other intelligence products in support of the joint force commander's decision-making process. It is a continuous process that includes defining the operational environment; describing the impact of the operational environment; evaluating the adversary; and determining adversary courses of action. Also called **JIPOE.** (JP 1-02. SOURCE: JP 2-01.3)

joint proponent. A Service, combatant command, or Joint Staff directorate assigned coordinating authority to lead the collaborative development and integration of joint capability with specific responsibilities designated by the Secretary of Defense. (Approved for inclusion in JP 1-02 with SecDef Memo 03748-09 as the source document.)

military assistance advisory group. A joint Service group, normally under the military command of a commander of a unified command and representing the Secretary of Defense, which primarily administers the US military assistance planning and programming in the host country. Also called **MAAG.** (Approved for incorporation into JP 1-02 with JP 3-22 as the source JP.)

Military Assistance Program. That portion of the US security assistance authorized by the Foreign Assistance Act of 1961, as amended, which provides defense articles and services to recipients on a nonreimbursable (grant) basis. Also called **MAP.** (JP 1-02. SOURCE: JP 3-22)

military civic action. The use of preponderantly indigenous military forces on projects useful to the local population at all levels in such fields as education, training, public works, agriculture, transportation, communications, health, sanitation, and others contributing to economic and social development, which would also serve to improve the standing of the military forces with the population. (US forces may at times advise or engage in military civic actions in overseas areas.) (JP 1-02. SOURCE: JP 3-57)

multinational operations. A collective term to describe military actions conducted by forces of two or more nations, usually undertaken within the structure of a coalition or alliance. (JP 1-02. SOURCE: JP 3-16)

nation assistance. Civil and/or military assistance rendered to a nation by foreign forces within that nation's territory during peacetime, crises or emergencies, or war based on agreements mutually concluded between nations. Nation assistance programs include, but are not limited to, security assistance, foreign internal defense, other Title 10, US Code programs, and activities performed on a reimbursable basis by Federal agencies or intergovernmental organizations. (JP 1-02. SOURCE: JP 3-0)

nongovernmental organization. A private, self-governing, not-for-profit organization dedicated to alleviating human suffering; and/or promoting education, health care, economic development, environmental protection, human rights, and conflict resolution; and/or encouraging the establishment of democratic institutions and civil society. Also called **NGO.** (JP 1-02. SOURCE: JP 3-08)

other government agency. Within the context of interagency coordination, a non Department of Defense agency of the United States Government. Also called **OGA.** (JP 1-02. SOURCE: JP 1)

paramilitary forces. Forces or groups distinct from the regular armed forces of any country, but resembling them in organization, equipment, training, or mission. (JP 1-02. SOURCE: JP 3-24)

peacekeeping. Military operations undertaken with the consent of all major parties to a dispute, designed to monitor and facilitate implementation of an agreement (cease fire, truce, or other such agreement) and support diplomatic efforts to reach a long-term political settlement. (JP 1-02. SOURCE: JP 3-07.3)

peace operations. A broad term that encompasses multiagency and multinational crisis response and limited contingency operations involving all instruments of national power with military missions to contain conflict, redress the peace, and shape the environment to support reconciliation and rebuilding and facilitate the transition to legitimate governance. Peace operations include peacekeeping, peace enforcement, peacemaking, peace building, and conflict prevention efforts. Also called **PO.** (JP 1-02. SOURCE: JP 3-07.3)

propaganda. Any form of adversary communication, especially of a biased or misleading nature, designed to influence the opinions, emotions, attitudes, or behavior of any group in order to benefit the sponsor, either directly or indirectly. (JP 1-02. SOURCE: JP 3-13.2)

psychological operations. Planned operations to convey selected information and indicators to foreign audiences to influence their emotions, motives, objective reasoning, and ultimately the behavior of foreign governments, organizations, groups, and individuals. The purpose of psychological operations is to induce or reinforce foreign attitudes and behavior favorable to the originator's objectives. Also called **PSYOP.** (JP 1-02. SOURCE: JP 3-13.2)

public diplomacy. 1. Those overt international public information activities of the United States Government designed to promote United States foreign policy objectives by seeking to understand, inform, and influence foreign audiences and opinion makers, and by broadening the dialogue between American citizens and institutions and their counterparts abroad. 2. In peace building, civilian agency efforts to promote an understanding of the reconstruction efforts, rule of law, and civic responsibility through public affairs and international public diplomacy operations. Its objective is to promote and sustain consent for peace building both within the host nation and externally in the region and in the larger international community. (JP 1-02. SOURCE: JP 3-07.3)

public information. Information of a military nature, the dissemination of which through public news media is not inconsistent with security, and the release of which is considered desirable or nonobjectionable to the responsible releasing agency. (JP 1-02. SOURCE: JP 3-13)

security assistance. Group of programs authorized by the Foreign Assistance Act of 1961, as amended, and the Arms Export Control Act of 1976, as amended, or other related statutes by which the United States provides defense articles, military training, and other defense-related services, by grant, loan, credit, or cash sales in furtherance of national policies and objectives. Security assistance is an element of security cooperation funded and authorized by Department of State to be administered by Department of Defense/Defense Security Cooperation Agency. Also called **SA.** (Approved for incorporation into JP 1-02.)

security assistance organization. None. (Approved for removal from JP 1-02.)

security cooperation. All Department of Defense interactions with foreign defense establishments to build defense relationships that promote specific US security interests, develop allied and friendly military capabilities for self-defense and multinational operations, and provide US forces with peacetime and contingency access to a host nation. Also called **SC.** (Approved for incorporation into JP 1-02.)

security cooperation organization. All Department of Defense elements located in a foreign country with assigned responsibilities for carrying out security assistance/cooperation management functions. It includes military assistance advisory groups, military missions and groups, offices of defense and military cooperation, liaison groups, and defense attaché personnel designated to perform security assistance/cooperation functions. Also called **SCO.** (Approved for inclusion in JP 1-02.)

security forces. Duly constituted military, paramilitary, police, and constabulary forces of a state. (Approved for inclusion JP 1-02.)

security force assistance. The Department of Defense activities that contribute to unified action by the US Government to support the development of the capacity and capability of foreign security forces and their supporting institutions. Also called **SFA.** (Approved for inclusion in JP 1-02.)

special operations forces. Those Active and Reserve Component forces of the Military Services designated by the Secretary of Defense and specifically organized, trained, and equipped to conduct and support special operations. Also called **SOF.** (JP 1-02. SOURCE: JP 3-05.1)

status-of-forces agreement. An agreement that defines the legal position of a visiting military force deployed in the territory of a friendly state. Agreements delineating the status of visiting military forces may be bilateral or multilateral. Provisions pertaining to the status of visiting forces may be set forth in a separate agreement, or they may form a part of a more comprehensive agreement. These provisions describe how the authorities of a visiting force may control members of that force and the amenability of the force or its members to the local law or to the authority of local officials. Also called **SOFA.** (JP 1-02. SOURCE: JP 3-16)

subversion. Action designed to undermine the military, economic, psychological, or political strength or morale of a governing authority. (JP 1-02. SOURCE: JP 3-24)

technical assistance. The providing of advice, assistance, and training pertaining to the installation, operation, and maintenance of equipment. (Approved for incorporation into JP 1-02 with JP 3-22 as the source JP.)

terrorism. The calculated use of unlawful violence or threat of unlawful violence to inculcate fear; intended to coerce or to intimidate governments or societies in the pursuit of goals that are generally political, religious, or ideological. (JP 1-02. SOURCE: JP 3-07.2)

unconventional warfare. A broad spectrum of military and paramilitary operations, normally of long duration, predominantly conducted through, with, or by indigenous or surrogate forces who are organized, trained, equipped, supported, and directed in varying degrees by an external source. It includes, but is not limited to, guerrilla warfare, subversion, sabotage, intelligence activities, and unconventional assisted recovery. Also called **UW.** (JP 1-02. SOURCE: JP 3-05)

unified action. The synchronization, coordination, and/or integration of the activities of governmental and nongovernmental entities with military operations to achieve unity of effort. (JP 1-02. SOURCE: JP 1)

JOINT DOCTRINE PUBLICATIONS HIERARCHY

JP 1
JOINT DOCTRINE
JP 1-0
PERSONNEL
JP 2-0
INTELLIGENCE
JP 3-0
OPERATIONS
JP 4-0
LOGISTICS
JP 5-0
PLANS
JP 6-0
COMMUNICATIONS SYSTEMS

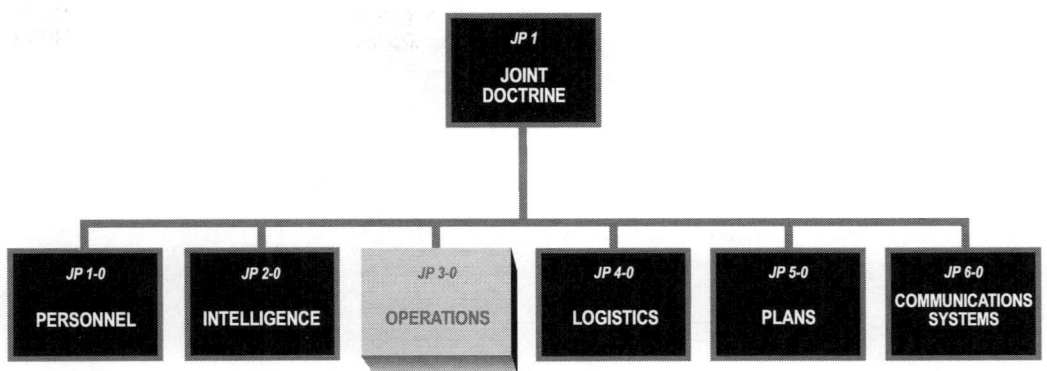

All joint publications are organized into a comprehensive hierarchy as shown in the chart above. **Joint Publication (JP) 3-22** is in the **Operations** series of joint doctrine publications. The diagram below illustrates an overview of the development process:

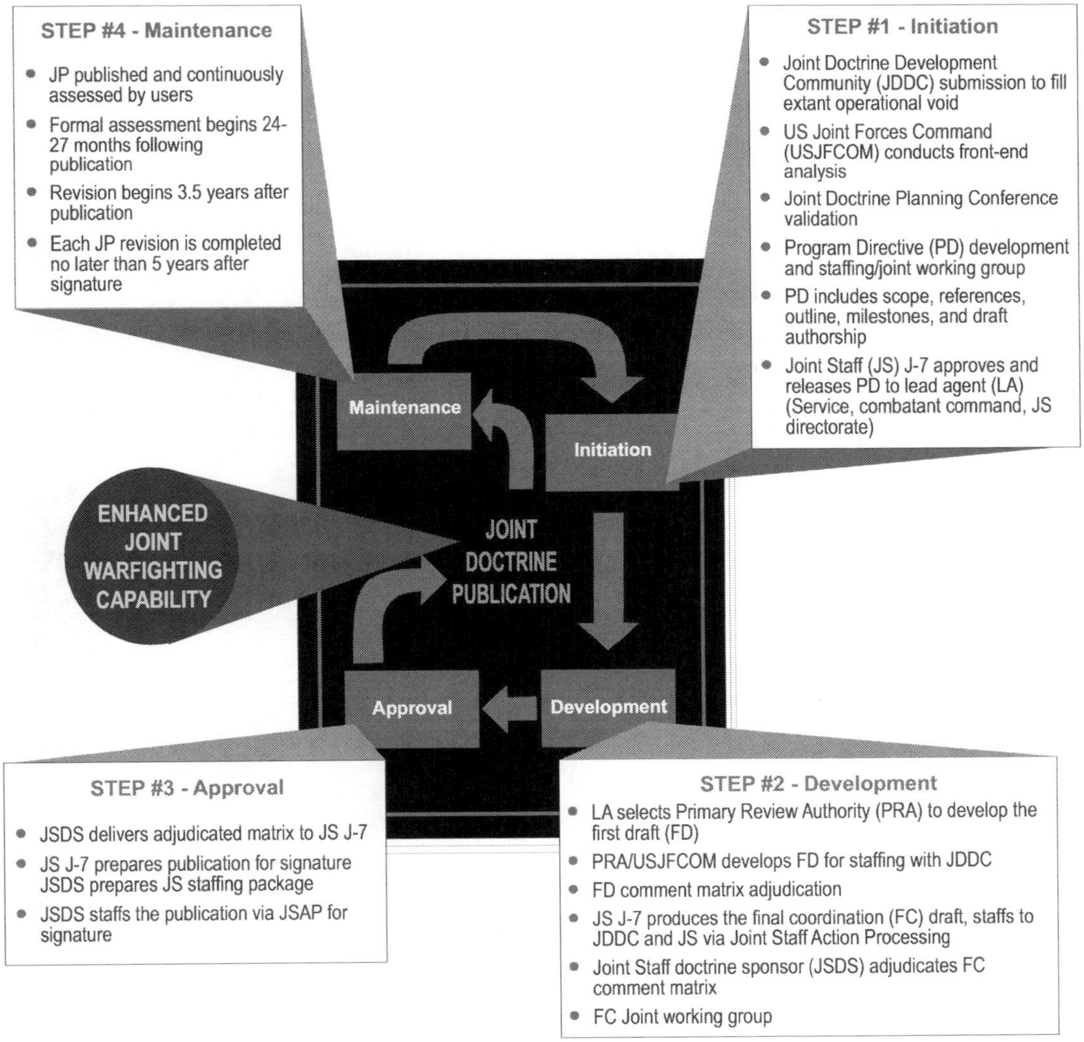

STEP #4 - Maintenance
JP published and continuously assessed by users
Formal assessment begins 24-27 months following publication
Revision begins 3.5 years after publication
Each JP revision is completed no later than 5 years after signature
STEP #1 - Initiation
Joint Doctrine Development Community (JDDC) submission to fill extant operational void
US Joint Forces Command (USJFCOM) conducts front-end analysis
Joint Doctrine Planning Conference validation
Program Directive (PD) development and staffing/joint working group
PD includes scope, references, outline, milestones, and draft authorship
Joint Staff (JS) J-7 approves and releases PD to lead agent (LA) (Service, combatant command, JS directorate)
Maintenance
Initiation
ENHANCED JOINT WARFIGHTING CAPABILITY
JOINT DOCTRINE PUBLICATION
Approval
Development
STEP #3 - Approval
JSDS delivers adjudicated matrix to JS J-7
JS J-7 prepares publication for signature JSDS prepares JS staffing package
JSDS staffs the publication via JSAP for signature
STEP #2 - Development
LA selects Primary Review Authority (PRA) to develop the first draft (FD)
PRA/USJFCOM develops FD for staffing with JDDC
FD comment matrix adjudication
JS J-7 produces the final coordination (FC) draft, staffs to JDDC and JS via Joint Staff Action Processing
Joint Staff doctrine sponsor (JSDS) adjudicates FC comment matrix
FC Joint working group

Made in the USA
San Bernardino, CA
13 October 2016